Generalised Physicochemical Model (PCM) for Wastewater Processes

Generalised Physicochemical Model (PCM) for Wastewater Processes

Edited by
Damien Batstone and Xavier Flores-Alsina

IWA PUBLISHING

Published by **IWA Publishing**
Unit 104–105, Export Building
1 Clove Crescent
London E14 2BA, UK
Telephone: +44 (0)20 7654 5500
Fax: +44 (0)20 7654 5555
Email: publications@iwap.co.uk
Web: www.iwapublishing.com

First published 2022

British Library Cataloguing in Publication Data
A CIP catalogue record for this book is available from the British Library

ISBN: 9781780409825 (Paperback)
ISBN: 9781780409832 (eBook)
ISBN: 9781780409849 (ePub)

This eBook was made Open Access in November 2022.

Content

Summary

Physicochemical reactions are those which occur spontaneously in conjunction with biochemical processes. These have received limited attention in biochemical reference models, though they have a strong impact on biochemical reactions, as well as directly describing processes such as gas transfer, precipitation, and metal species oxidation and reduction. An effective physicochemical model is critical to plant-wide modelling, and particularly simulation of phosphorous, both at a unit, and plant-wide level. The range of physicochemical reactions in complex, but very well supported by fundamental aquatic chemistry theory, and a common approach is suitable in a plant-wide context, or even outside these boundaries, upstream to the sewer system, and downstream to the receiving catchment.

In this report, the theory and practice of aquatic chemistry applied to wastewater processes is described. Chapter 1 describes motivation for a generalised model, as well as the state of art in published IWA models. Chapter 2 describes liquid-phase equilibria, with acid–base equilibria incorporating the range of compounds generally found in wastewater treatment, and including non-acid–base ion pairing. Inclusion of non-ideality using (generally Davies) is necessary at most wastewater strengths. Formulation of an implicit algebraic equation set using the Tableau approach is described, and we describe dimensional simplification of the resulting equation set, and solution using an implicit algebraic solver. The liquid phase equilibria are the core of the generalised physicochemical model, and enables effective species activity determination for all other reactions.

Chapter 3 describes precipitation, which is the formation of particulate compounds from soluble species, which are generally (but not always) a combination of charged species to form a neutral precipitate. Common precipitates are formed by calcium, magnesium, iron, and aluminium carbonates, phosphates, hydroxides, and sulfides. Most precipitates have a fixed stoichiometry, and their formation is modelled by a kinetic expression of the supersaturation. However, in the mainline, phosphate removal with iron or aluminium is a multi-step process of formation of the aluminium or iron hydroxide particle (HAO or HFO, respectively), followed by adsorption of phosphate on available HAO or HFO sites. This results in a variable stoichiometry, which utilises an alternative approach.

Chapter 4 describes the theory of gas–liquid transfer, including development of two-film theory, and its application to bidirectional gas transfer in wastewater processes. This extends to the determination of gas transfer coefficients, and the basic theory, including fundamentals of neutral species activity and temperature correction are described in conjunction with aeration theory, which has been extremely extensively developed.

Chapter 5 describes spontaneous chemical oxidation and reduction, with a focus on iron oxidation and reduction, due to its relevance to phosphate chemistry. These reactions are actually generally catalysed by biomass, but are assumed not to contribute to microbial yield. Oxidation and reduction potential (ORP) is described, with iron being oxidised under aerobic or anoxic conditions, and being reduced under anaerobic conditions. A second-order relationship is proposed, with oxygen or nitrate as electron acceptor under aerobic conditions (for iron oxidation) and sulfide or hydrogen as electron donor under anaerobic conditions.

Chapter 6 describes application of advanced physicochemical models in a range of plant-wide applications, both domestic and industrial. This particularly focuses on effective prediction of phosphorus dynamics, but also extends, e.g. to mineral precipitation in biofilms.

The description of theory and application provided here is supported by a reference implementation in a plant-wide context in Matlab C format. A link to the repository is provided in the Appendix.

This book is dedicated to Professor George Ekama, University of Cape Town, who has educated and inspired generations of scientists and engineers worldwide, and who has been a driving force behind the development of integrated chemical, physical and biological wastewater treatment process models.

Acknowledgements

This book is the result of extensive and highly productive collaboration, across academia and industry. The taskgroup was established based on a workshop at WWTMod2010 (2nd international IWA/WEF wastewater treatment (WWT) modelling), at Mont Sainte Anne, Quebec, Canada. While this conference series is a team effort, we particularly recognise the leadership of Peter Vanrolleghem from Université Laval in re-initiating this wonderful series of symposiums, as well as his enthusiastic participation in, and support of the taskgroup activities. We have maintained this collaboration and presented outcomes at later WWTMod (subsequently WWRMod) conferences, as well as via the IWA Watermatex conferences, and this connection has been vital to the continued collaboration.

Authoring this book has been a long endeavour, and we thank the editorial and book team of IWA, particularly Mark Hammond for his patient, persistent, and polite support over the many years.

Aquatic chemistry is a complex topic, particularly when applied in conjunction with biochemical processes such as wastewater treatment. Many of us would have found it far more difficult to gain a basic understanding, and we acknowledge in particular the contributions of George Ekama from University of Cape Town, and Chris Brouckaert from University of KwaZulu-Natal for their extensive work in explaining the complexities of combined biological and chemical reactions in wastewater environments.

doi: 10.2166/9781780409832_0001

Chapter 1
Introduction

Taskgroup

1.1 PHYSICOCHEMICAL REACTIONS

Physicochemical reactions refer to those that occur spontaneously, without the need for a biological mediator. This is in contrast with biochemical reactions, which require the presence of a biological catalyst (normally microbial biomass, or enzymes produced by microbes). The major proportions of physicochemical reactions in wastewater are spontaneous liquid-phase reactions, either interacting with other solutes in a water matrix, or with gas and solid phases. These *largely* do not involve molecular reduction or oxidation. The major classes of reactions (and time scales) are provided in Table 1.1 Examples of some of the compounds and reactions are shown in Figure 1.1.

Physicochemical reactions are highly important to all elements of the urban, industrial and natural water cycle, from water collection to impact on the receiving environment. They act as the basis for principal unit operations such as pH adjustment, coagulation-chemical precipitation, gas stripping, and chemical oxidation and reduction, and have a strong impact on both biological and physical processes (such as separation). Wastewater quality and operability measures such as dissolved oxygen and pH are intimately linked to physicochemical reactions. An appropriate representation of physicochemical processes is critical to effectively simulate water and wastewater treatment processes. The generalised physicochemical modelling (PCM) framework presented in this report, focuses on application in the wastewater context, but has broad applicability to drinking water systems and the natural environment (including saline-receiving environments).

1.2 PCM IN WASTEWATER TREATMENT PROCESS MODELS

Process modelling is widely used in water and wastewater treatment design, operation, diagnosis, and research (Gernaey *et al.*, 2014; Henze *et al.*, 2000; Jeppsson *et al.*, 2013). For wastewater treatment, this is supported by a diverse range of commercial software and reference models, including the IWA models, which comprise the activated sludge series (ASM1, ASM2/ASM2d, ASM3) (Henze *et al.*, 2000), focusing mainly on nutrient removal, the Anaerobic Digestion Model No. 1 (Batstone *et al.*, 2002a) and the River Water Quality Model No. 1 (Reichert *et al.*, 2001). The level of PCM in IWA biochemical models is generally rudimentary, at a basic level needed to support the core biological model (Batstone *et al.*, 2012) (Table 1.2). Specifically, in the liquid phase, the ASM series represents

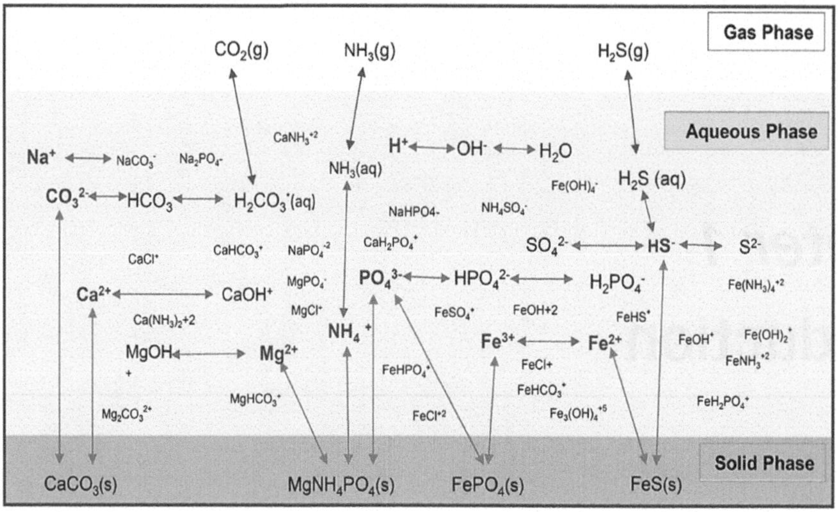

Figure 1.1 Physicochemical reactions and species commonly occurring in wastewater processes (Source: from Kazadi Mbamba (2016), used with permission).

acid–base as an alkalinity state, which is depleted through nitrification, and regenerated through denitrification. The ADM explicitly calculates pH through a charge balance, though without activity correction and ion pairing (see later sections). Both ASM/ADM contain liquid film-controlled gas transfer. The ASM series either considers the atmosphere as an infinite source/sink of oxygen at atmospheric pressure (with a gas transfer coefficient k_La), or set dissolved oxygen to a constant (Henze *et al.*, 2000). The ASM2d includes phosphate complexation through an empirical relationship. A more extensive discussion of the limitations of PCM in the IWA models is included in the scoping paper for this technical report (Batstone *et al.*, 2012).

The IWA models had an appropriate capability for the scope requirements of the time, largely limited to biological nitrogen removal (and associated assessment of aeration requirements and basic mainline plant simulation including sludge production), or phosphorus removal for the ASM2/2d, and COD removal in the ADM1. However, particularly the need to predict phosphorus transformations, and in the context of plant-wide modelling for system optimisation, and resource recovery, has exceeded these basic specifications. Limitations in the IWA models have been recognised in literature reviews

Table 1.1 Physicochemical reaction classes.

Type of reaction	Key example	Importance	Speed (time constants)
Liquid-phase ion reactions	Acid–base reactions Ion pairing	Determines pH, acid–base species concentration	Very fast ($<s^{-1}$)
Liquid–solid-phase reactions	Ionic precipitation and redissolution	Determines formation rate of solids	Medium (min^{-1} h^{-1})
Liquid-gas-phase reactions	Gas–liquid transfer	Determines gas transfer rate for aeration and gas stripping	Medium–fast (s^{-1} min^{-1})
Chemical oxidation and reduction	Solute reduction and oxidation with H_2 and O_2	Chemical oxidation and reduction	Fast (s^{-1} min^{-1})

Table 1.2 Current approaches to physicochemical systems in IWA models.

Model	Acid–base	Gas–liquid	Solid–liquid
ASM1/ASM3[1]	Alkalinity state	Liquid-film controlled (k_La)	None
ASM2/ASM2d[1]	Alkalinity state	Liquid-film controlled (k_La)	Empirical P complex
ADM1[2]	pH calculated	Liquid-film controlled (k_La)	None
RWQM1[3]	pH calculated	Liquid-film controlled (k_La)	Empirical

[1]Gas–liquid transfer included from atmosphere by k_La or oxygen set.

(Batstone *et al.*, 2012, 2015), and through specific improvements to the ASM2D (Serralta *et al.*, 2004) and ADM1 (Zhang *et al.*, 2015). Advanced and comprehensive chemistry models have been proposed (Musvoto *et al.*, 2000a, 2000b; Sötemann *et al.*, 2005), and these principles are increasingly being applied to IWA models, particularly in plant-wide models (Flores-Alsina *et al.*, 2015; Lizarralde *et al.*, 2015).

The majority of applied wastewater process modelling is done in commercial software, which, while broadly compatible with IWA states, kinetics, and units, has evolved significantly from the IWA models in response to user needs. This has included implementation of comprehensive chemistry models incorporating detailed physicochemistry, with commonly, plant-wide pH, complexation and precipitation, and advanced gas transfer (Fairlamb *et al.*, 2003; Hauduc *et al.*, 2019). It is fair to say that commercial software has generally led academic research into applied, practical, and plant-wide implementation of advanced chemistry modelling, highlighting the practical importance of physicochemical processes.

The trend of incorporating advanced chemistry has accelerated since this Task Group was established in 2012 (Batstone *et al.*, 2012) (see e.g. Chapter 6). Most commercial and advanced research plant-wide simulators now, including pH simulation throughout the water line, incorporating other aspects of complex chemistry (including metal and phosphate reactions). One of the goals of this report is to explain the common theory and basis for these models.

1.3 AQUATIC CHEMISTRY MODELLING

Geochemistry and aquatic chemistry in natural waters is a very well-established scientific field, to which fundamental classical thermodynamics apply (Stumm & Morgan, 1996). This provided a strong basis for the PCM framework. The key references are provided here, and utilised in the following chapters (Morel & Hering, 1993; Stumm & Morgan, 1996). The two broadly used (open source) software tools for modelling aquatic and geochemistry are PHREEQC (Parkhurst & Appelo, 2013) (US Geological Survey) and MINTEQ(A2) (Allison *et al.*, 1991) (US EPA). There are a number of other packages available (EQ3/6, WATEQ4F), with varying levels of maintenance and capability. In line with the long time scales of natural systems, most of these packages are designed to solve for the equilibrium condition with respect to all phases and redox, though kinetics can be included. The speciation models are generally very sophisticated, including complex ion chemistry and a detailed consideration of non-ideality. PHREEQC has the highest technical capability (based on our review and external) (Merkel *et al.*, 2008), generally providing comprehensive activity corrections for all waters, and with the inclusion of the complex Pitzer ion activity model, allowing simulation of saline environments. In contrast, MINTEQ(A2) uses simpler activity corrections only valid at lower salinity levels (see below) but has a broader compound database and the ability to user customise.

These generalised packages have applicability to wastewater process modelling, and they have been applied to biological models, with PHREEQC as an example, either by linking the aquatic chemistry simulator to the biological model (Vaneeckhaute *et al.*, 2018), or by solving the biological model in the PHREEQC dynamic solver (Huber *et al.*, 2017). This brings a high level of PCM capability to wastewater process modelling, but at the expense of a somewhat monolithic aquatic chemistry structure, with reduced solver times, and reduced transparency. Moreover, because the underlying assumptions of the base models (i.e., all reactions achieve equilibrium) does not always apply to wastewater, base models have to be corrected or redox conditions carefully managed to prevent erroneous outputs. This has been addressed by for example PHREEQC model size reduction (Vaneeckhaute *et al.*, 2018), but still carries substantial computational overhead.

1.4 NEEDS-ANALYSIS, AIMS OF THIS BOOK

Plant-wide process modelling is becoming essential to the field of wastewater treatment process development, design, and application (Jeppsson *et al.*, 2013). Benchmarking, for example, requires years of simulated time under dynamic conditions (time steps on the order of 15 min), and due to the phenomena of harmonic interactions, design prototyping (in industrial applications) should apply the same principles. That is, steady state or short-term dynamic modelling (~months) cannot expose plant-wide harmonic interactions that may occur over long time scales between influent, controllers, internal recycles, and sludge (concentrate recycles), which may represent substantial design flaws. The mixture of dynamic time scales (seconds to days), presence of controllers (and inclusion of actuators and sensor noise/delays), and the dynamic nature of the influent, impose a computationally intensive problem. Therefore, any additional overhead from the PCM should be minimised.

However, plant-wide commonality for the PCM is essential. This not only enables process-wide state compatibility, which is sensible, since the states are commonly present throughout the process, but also allows for iron–sulfur–phosphorus modelling, which is linked across activated sludge and anaerobic digester modelling, and requires an integrated set of biochemical and physicochemical transformations (Flores-Alsina *et al.*, 2016).

A number of approaches are appropriate, including the use of dedicated simplified aquatic chemistry models (such as used in the ADM1) (Batstone *et al.*, 2002a), or application of tools from other fields (such as PHREEQC) (Vaneeckhaute *et al.*, 2018). However, there is a need for a high-capability, generalised plant-wide physicochemical model for wastewater systems, given the range of systems to be analysed. This will be especially important with the emerging importance of resource recovery concepts within the urban and industrial wastewater context, with a combination of physicochemical and biological technologies being proposed and strongly interacting within the plant-wide context. Having a consistent PCM platform for assessment, design and operations, will be essential to support the adoption of such resource recovery concepts.

This report aims to present theory and principles behind these models, provide information for appropriate model selection or reduction, provide a reference implementation, and enable application across water cycle modelling, focusing on wastewater process modelling.

doi: 10.2166/9781780409832_0005

Chapter 2

Liquid (aqueous)-phase chemistry

Damien J. Batstone[1], Xavier Flores-Alsina[2], Christian Kazadi Mbamba[1] and Stephan Tait[3]

[1]Australian Centre for Water and Environmental Biotechnology, The University of Queensland, St Lucia, Brisbane, QLD 4072, Australia
[2]PROSYS Research Center, Department of Chemical and Biochemical Engineering, Technical University of Denmark, Building 229, DK-2800, Kgs. Lyngby, Denmark
[3]Centre for Agricultural Engineering, University of Southern Queensland, Toowoomba 4350, Australia

2.1 EQUILIBRIUM CHEMISTRY AND AQUATIC CHEMISTRY

This chapter presents thermodynamics and equilibria chemistry from the perspective of aquatic chemistry (i.e., water as solvent). There is a broader discussion of fundamentals and terms (including non-aqueous systems) in the key reference for this chapter (Stumm & Morgan, 1996). A more wastewater targeted discussion (particularly of component and species in the wastewater context) can be found in Brouckaert *et al.* (2021, 2022).

Aquatic chemistry modelling depends heavily on equilibrium theory, as commonly, liquid (solute)-phase reactions are assumed to be at (reversible) equilibrium within the timeframes applicable to wastewater systems. This means the equilibrium condition can be efficiently solved through algebraic equations, given a basic specification, as discussed below. While it is also possible to solve such a system using differential equations, a differential equation approach becomes unmanageable for complex systems as further discussed in Section 2.3, and even then, basic formulation still follows equilibrium principles, with conversion to differential equations during implementation.

Given equilibrium conditions apply, the concentrations of reactants and products are at equilibrium. This equilibrium can be related to a reference standard state, which is an ideal mixture of water (and solutes) at 298.15 K, 1.013 bar. The effective equilibrium changes as the system state moves from this reference condition, particularly as the mixture becomes non-ideal as noted below. Correction of temperature is noted below, and aqueous solutions are generally assumed incompressible, which means that pressure has minimal effect on solution state.

The formal definition of an ideal mixture is that enthalpy of mixing is zero (Matthews, 2000). In an ideal mixture, a number of properties, including vapour pressure, heat capacity, density, and reaction equilibria are algebraically related to the molar concentration (or fraction) of individual constituents. Ideality applies practically for aqueous solutions, only at infinite dilution. That is, the system is dilute enough to be considered ideal. As the concentration of solutes increases, the reactivity of a solute is expressed in the form of the activity, which is related to molar concentration by its respective activity coefficient γ_i. Specifically, for species i (S_i), activity $\{S_i\}$ (often also termed a_i) is calculated from molar concentration $[S_i]$ (often also termed m_i) as follows:

$$\{S_i\} = \gamma_i [S_i] \tag{2.1}$$

Note that in this book, molar (mole L^{-1}) units are used rather than molal (mole kg^{-1}), but the concepts also apply to molal units. By convention both the activity $\{S_i\}$ and activity coefficient γ_i are non-dimensional and hence the right-hand side of equation (2.1) is implicitly divided by the standard concentration (1 mole L^{-1} for molar units) (Cohen *et al.*, 2007).

At infinite dilution, $\gamma_i = 1$, as the system is an ideal mixture at the reference condition. With movement away from infinite dilution, activity is determined by applying a value γ_i estimated depending on the ionic strength of the water, as discussed later in this chapter. Here, activity refers to the 'effective concentration' of a solute (the species i) dissolved in a solvent (the water or aqueous phase, in this case). The equivalent for activity in gaseous mixtures at elevated pressure is fugacity.

To illustrate equilibrium concepts, consider the generalised reversible reaction between reactants A and B and products C and D, with the forward and back reactions occurring simultaneously:

$$aA + bB \leftrightarrow cC + dD \tag{2.2}$$

At equilibrium (for a given temperature and solvent condition), the forward and back reactions are balanced (no net change) and activities of the reactants and products ([A]–[D]) satisfy an equilibrium relationship.

$$K = \frac{\{C\}^c \{D\}^d}{\{A\}^a \{B\}^b} \tag{2.3}$$

with an equilibrium constant K.

The K value is a fundamental value, and is related to the free energy of reaction (ΔG^0_{rxn}) at normal temperature and pressure (i.e., standard state, 298°K, 1.013 bar, and pure water for solutions) as follows:

$$\Delta G^0_{\text{rxn}} = -RT \ln K \tag{2.4}$$

Hence, the K value is only dependent on temperature (and pressure), rather than a contestable (i.e., needs to be fitted from data) coefficient or parameter, and can be found at reference conditions in reference texts (or determined from ΔG^0 values).

Due to the underlying relationship of free energy with enthalpy and entropy ($\Delta G^0 = \Delta H^0 + T\Delta S^0$), if enthalpy ($\Delta H^0$) is assumed to be constant over the temperature range considered (T_1–T_2), the K value can be corrected for temperature using the van't Hoff equation (Puigdomenech *et al.*, 1997; Stumm & Morgan, 1996) as follows:

$$\ln \frac{K_1}{K_2} = \frac{\Delta H^0_{\text{rxn}}}{R} \left(\frac{1}{T_1} - \frac{1}{T_2} \right) \tag{2.5}$$

If enthalpy is not constant with temperature, a correction can be addressed through by inclusion of heat capacity c_p across the range T_1–T_2 (Puigdomenech *et al.*, 1997), but this is not generally necessary in the wastewater context (Tait *et al.*, 2012).

2.1.1 Acid–base equilibria

In aquatic chemistry, water is the solvent, and is not included in the equilibrium calculation (or rather, implicitly included in the ΔG^0_f values of the reactants). For example, the reaction $H_2O \leftrightarrow H^+ + OH^-$ has an equilibrium expression

$$K_w = \{OH^-\}\{H^+\} = 10^{-14} \tag{2.6}$$

Acid–base reactions are generally (in practical terms – see Chapter 1) at equilibrium, and the generalised acid–base for acid (HA) to base (A^-) reaction is written in the direction:

$$HA \leftrightarrow A^- + H^+ \tag{2.7}$$

Note that the base can be charged or uncharged (with the reciprocal acid having a valency 1 higher than the base). The generalised equilibrium equation is

$$K_A = \frac{\{A^-\}\{H^+\}}{\{HA\}} \tag{2.8}$$

Here, K_A is commonly termed the acidity constant, and its mathematical variant pK_A ($=-\log_{10}(K_A)$) can typically be sourced from reference texts.

Many compounds (e.g., ammonia, organic acids) have only a single acid/base pair in the normal pH range of aquatic chemistry, but many compounds (sulfide, inorganic carbon, phosphates) have two or more individual acids or bases, with an equilibrium expression for each acid–base pair (e.g., inorganic carbon has an expression each for the reactions $CO_{2,aq} + H_2O \leftrightarrow HCO_3^- + H^+$ and $HCO_3^- \leftrightarrow CO_3^{2-} + H^+$.

2.1.1.1 Components and species

As described further below, in Section 2.1.3, when taking an algebraic approach to calculating equilibrium in the aqueous liquid phase, a specific number of independent solution constituents (components) needs to be specified to fully describe the total molar composition of the liquid phase. These define the solution state, from which individual concentrations of species can be algebraically calculated. For the purposes of acid–base equilibria, components can be arbitrarily defined, as long as they are mathematically linearly independent. However, when physicochemical modelling is used in a wastewater environment, it is convenient to take a specific approach where the components are defined as the combined state of a real recognisable/measurable compound (e.g., combining both acid and base species into a total component state, such as HS- and H_2S species being combined into a component, total dissolved sulfide). This section provides a generalised discussion of the concepts of components and species, applicable to wastewater applications. However, the approach provided here is generalised and high-level, and a more targeted, wastewater focused discussion, including incorporation with biology and specific measures such as alkalinity is provided in Brouckaert *et al.* (2021, 2022).

Dynamic process modelling of biological processes generally computes the component concentration as a time-dependent differential state variable from the mass conservation equation. Most analytical methods also measure, or report the component concentration. For acid–base reactions, this combined concentration can be expressed as

$$[A_{tot}] = [A^-] + [HA] \tag{2.9}$$

As discussed below, the combined concentration $[A_{tot}]$ is a system closure input (component) which reduces degrees of freedom, and enables calculation of the acid–base system algebraically (species) using equilibrium relationships. For the remainder of this section, only ideal conditions are discussed, with non-dilute conditions addressed later in the chapter. Note that non-ideality generally must be applied. It is simply convenient to discuss basic equation formulation applying an ideal assumption.

In an ideal mixture, equation (2.9) can be combined with equation (2.8) (with $\gamma_i = 1$) to calculate concentrations of individual species from the component concentration:

$$[HA] = \frac{[H^+][A_{tot}]}{K_a + [H^+]} \quad \text{infinite dilution, standard state} \tag{2.10}$$

$$\left[\mathrm{A}^-\right] = \frac{K_a\left[A_{tot}\right]}{K_a + [\mathrm{H}^+]} \quad \text{infinite dilution, standard state} \tag{2.11}$$

For each acid–base pair, there are two unknowns (concentrations of the acid and base species), and two equations (the species contribution balance equation (2.9), and the acid–base equilibria equation (2.7)). If the component state is known (e.g., from dynamic mass balance model), the only remaining unknown is the proton concentration [H+]. See equation (2.3.2) which indicates how equations (2.10–2.11) must be modified by the use of K_A' or K_A'' under non-ideal conditions.

Proton activity is measured and expressed on the negative $\log_{10}$ scale termed pH (pH $= -\log\{H^+\}$). Because K_a is often represented on a negative $\log_{10}$ scale $pK_a = -\log(K_a)$ (as above), reference to equation (2.10) can show that at pH $= pK_a$, 50% of the combined acid + base is in the acid or the base form, and similarly at pH $= pK_a + 1$, 90% is in the base form, and at pH $= pK_a + 2$, 99% is in the base form.

2.1.1.2 Closing acid–base system

In the above, in an ideal mixture (i.e., $\{A\} = [A]$), there is one unknown, which is the proton concentration.

This can be resolved in two ways;

(i) Charge balance. This applies the condition of electroneutrality, where the total concentration of positive charges must equal the total concentration of negative charges so that the overall net charge is zero. That is, in general,

$$0 = \sum_i z_i\left[S_i^{z_i}\right] - \left[\mathrm{OH}^-\right] + \left[\mathrm{H}^+\right] \tag{2.12}$$

where $S_i^{z_i}$ is the molar concentration of ionic species i with charge z_i. Given the water acid–base equilibrium equation (equation 2.6), the system is fully specified. Note that equation (2.12) is in concentration units rather than activities even for a non-ideal solution.

(ii) Proton balance approach. This applies the principle that the total free and bound protons present in a system must be fixed by the components in that system; that is adding $[A_{tot}]$ as the acid HA will increase the proton inventory proportionally, while adding it as A^- (e.g., as a sodium salt) will not increase the proton inventory. That is, in general,

$$0 = \sum_i y\left[H_{y_i}S_i\right]_{added} - \sum_i y\left[H_{y_i}S_i\right]_{eq} + \left[\mathrm{OH}^-\right] - \left[\mathrm{H}^+\right] \tag{2.13}$$

where $y\left[H_{y_i}S_i\right]_{added}$ is the cumulative amount of protonated acid added to solution, with y protons available, and $\sum_i y\left[H_{y_i}S_i\right]_{eq}$ is the equilibrium concentration of the protonated acid species. S_i may be negatively or positively charged. While the two approaches seem radically different, both result in the same numerical outcome, and both are equally applicable to wastewater problems. In both cases (except for synthetic solutions), it is impossible to know exactly what mixture of acids or bases are present, and a solution background is generally applied, which sets the net background valency for the charge balance, or background proton state for the proton balance approach. This is generally done by solving for known pH at the initial condition, and leaving the charge or proton balance state constant in future simulations.

Taking a simple system to demonstrate equivalence of the two approaches, for the acid HA and base A^-, where 1 mole of HA is added to 1 L pure water – 1 M (mole L^{-1}) solution. While noting a 1 M solution is far from dilute, applying an ideal solution approach for illustration ($\{A\} = [A]$) results in the following equations:

$$[A_{tot}] = [A^-] + [HA] = 1\,M \tag{2.14}$$

$$K_A = \frac{[A^-][H^+]}{[HA]} \tag{2.15}$$

$$K_w = [OH^-][H^+] = 10^{-14} \tag{2.16}$$

And either

$$0 = -[A^-] - [OH^-] + [H^+]\ (\text{charge balance}) \tag{2.17}$$

or

$$0 = [HA]_{added} - [HA] + [OH^-] - [H^+]\ (\text{proton balance}) \tag{2.18}$$

where $[HA]_{added} = [A_{tot}] = 1$ M.

It is evident that equation (2.18) is a linear combination of equations (2.14 and 2.17). This can be reformulated analytically as a cubic polynomial (in $[H^+]$), and solving this reveals that $pH = pK_a/2$. Adding instead 1 M of the base (as for example NaA) would result in $[A_{tot}] = 1\,M$, $[HA]_{added} = 0\,M$ and a solved $pH = 7 + pK_a/2$. Adding equimolar concentrations of acid and base results in $pH = pK_a$.

In a wastewater (or even drinking water) context, there are a large number of acids and bases, and the system rapidly becomes unmanageable analytically, since each acid or base adds to the polynomial order and analytical reformulation effort. If the net components are known fully, or partially, and a background proton or charge balance condition applied, the concentration of protonated and unprotonated species, including $[H^+]$ and $[OH^-]$ can be determined through the implicit algebraic equation set. Approaches to solving this algebraic set of equations (or a reduced set of differential equations) are further discussed below.

2.1.2 Ion pairing equilibria

Acid–base equilibria are a subset of generalised liquid-phase equilibria, specifically involving association and dissociation of solute species with protons and hydroxide ions. In general, solute species (including acids or bases) can associate with other solute species to form ion pairs. For example, the cation Na^+ can associate or pair with the anion HCO_3^- according to the equilibrium equation ($NaHCO_{3,aq} \Leftrightarrow Na^+ + HCO_3^-$), described by the equilibrium relationship:

$$K_{NaHCO_3} = \frac{\{Na^+\}\{HCO_3^-\}}{\{NaHCO_3\}} \quad K_{NaHCO_3} = 1.77\,M \tag{2.19}$$

It is important to note that this $NaHCO_{3,aq}$ is not a solid salt $NaHCO_{3,s}$, but instead a liquid-phase ion pair species.

Ion pairing reactions have an important impact on acid–base equilibria, gas transfer, and precipitation, as paired ions (e.g., $NaHCO_{3,aq}$) reduce the concentration (and thus the activity) of free ion species (e.g., Na+ or HCO_3^-) at equilibrium. The effect on the charge or proton balance is nil (charges cancel out, and protons are preserved via the relevant species), though consistency is important.

Ion pairing is important at higher water or wastewater strengths, and it has been identified as being equally important to general non-ideality (i.e., activity vs concentration) in landfill leachate

(Tait *et al.*, 2009) and anaerobic digesters (Solon *et al.*, 2015). Including ion pairing is particularly important in the presence of di-valent and tri-valent ions, since multiple paired species can exist (with varying ionic charge) (Flores-Alsina *et al.*, 2015). As for acid–base equilibria, each ion pair considered will necessarily add an additional equation to the system of general algebraic equations, so that the model system is fully specified. A very large number of pairs can exist within a wastewater system, and while the K values and individual equations are typically known, this adds substantial complexity to the system of equations to solve. The number of potential pairs can be eliminated by evaluation of the equilibrium K_i values (e.g., from the PHREEQC database (Parkhurst & Appelo, 1999). As an example, $NaCl_{aq}$ may be excluded except under saline conditions as its K_i value is relatively high at $10^{1.582}$ M ($pK_i = -1.582$).

2.1.3 Large systems – the Tableau approach

As discussed in Section 2.1.2, with acid–base and ion pairing relationships defining equilibrium variables (species), a number of components or closure variables are required to fully define the system. Selection of components can be somewhat arbitrary, but components must be mathematically independent, and must be able to be expressed mathematically as a linear combination of the concentrations of contributing species.

The number of components needing definition can be precisely determined as (Morel & Hering, 1993):

$$N_C = N_S - N_R \tag{2.20}$$

where N_C is the number of components needing definition, N_S is the number of species to be determined, and N_R is the number of independent equilibrium reactions/relationships. In dynamic modelling, it is usually preferred to source these from differential states, as discussed above. It would be possible though to use a different set of components, as long as they act effectively to close the algebraic equation set. An example method which is commonly used is to express all components as the constituents of acid/base sets with the highest positive or negative charge (e.g., PO_4^{3-}, CO_3^{2-}, Ca^{2+}, Na^+, NH_4^+). This is convenient conceptually, but can cause numerical issues, with multi-valent acids and bases (e.g., PO_4^{3-}, CO_3^{2-}) having very low values at neutral pH. Water is always a component, but is removed from the equation set and mass balance and is instead included implicitly. Either total proton concentration or total net charge is generally a component, expressed via a charge or proton balance, as above.

Another consequence of this is that additional reactions (adding to N_R) do not necessarily increase N_C if it only adds a single species to the entire problem.

Systems can be very complex in terms of number of species and reactions, but quite minimal with respect to number of components. For example, the recommended component/species system for plant-wide modelling (the default PCM for plant-wide modelling) includes only 20 components (Kazadi Mbamba, 2016) which are, in order of elemental symbol, total trivalent aluminium (S_{Al}), total inorganic carbon (S_{IC}), total acetate (S_{ac}), total propionate (S_{pr}), total butyrate (S_{bu}), total valerate (S_{va}), total dissolved calcium (S_{Ca}), total chloride (S_{Cl}), total dissolved ferrous (S_{Fe2}), total dissolved ferric (S_{Fe3}), protons (implicit), total dissolved magnesium (S_{Mg}), total sodium (S_{Na}), total potassium (S_K), total phosphate (S_{IP}), total dissolved ammoniacal nitrogen (S_{IN}), total nitrate (S_{NO3}), total nitrite (S_{NO2}), total dissolved sulfide (S_{IS}), total dissolved sulfate (S_{SO4}), but includes 118 corresponding species, and hence requires 98 independent equilibrium reactions/relationships (Solon *et al.*, 2017).

It is convenient to represent equilibrium systems, consisting of acid–base and ion-pairing species in a tabular format (commonly referred to as a Tableau) (Morel & Hering, 1993; Stumm & Morgan, 1996). This provides a compact format, as well as ensuring that the basic independence and representative requirements for components are met. The table consists of components in the columns, and species in the rows.

Table 2.1 Tableau representation of components and algebraic species for a simple system.

Species $(i)\downarrow$	Components→ (TOT_j)	TOTH* 1	S_{IC} 2	S_{AC} 3	S_{Na} 4	S_{Cl} 5	S_{IN} 6	TOT (H_2O)	$-\log(K_i)$
1	H^+	1							
2	CO_3^{2-}		1						
3	Ac^-			1					
4	Na^+				1				
5	Cl^-					1			
6	NH_4^+						1		
	(H_2O)							1	
7	OH^-	−1							14.18
8	HCO_3^-	1	1						10.33
9	$CO_2(aq)/H_2CO_3$	2	1					−1	16.68
10	HAc	1		1					4.76
11	NH_3	−1					1		−9.24
12	$NaCO_3^-$		1		1				1.27
13	$NaHCO_{3(aq)}$	1	1		1				10.03
14	$NaAc_{(aq)}$			1	1				−0.12
15	$NaOH_{(aq)}$	−1			1				−13.9

Note: NaCl, Na_2CO_3, $NH_4CO_3^-$, $NH_4HCO_{3(aq)}$, $NH_4Ac_{(aq)}$, $NH_4OH_{(aq)}$ and $NH_4Cl_{(aq)}$ do not form in substantial amounts and are therefore omitted. $NaCl_{aq}$ may be included for high salinity systems with a $pK_i = -1.582$. Note that where the base is the dominant species ($i < N_c$; $i < 6$), $K_i = 1/K_a$. Where the acid is the dominant species (e.g., NH_4^+), $K_i = K_a$. Equilibrium coefficients are taken from the PHREEQC database with reaction stoichiometry as shown in the table (reactants −, products +).

The entries in the Tableau represent the extent to which that species contributes to a component. For example, the component total inorganic carbon is made up of dissolved CO_2, HCO_3, or CO_3^{2-} and all species that contain them. It is convenient (but not necessary) to identify which of the species is related to the equilibrium, and which is related to the mass balance (e.g., by influencing rates of reactions in the dynamic equation set), even though all are resolved through an implicit algebraic equation set. These are the first N_C species in the table. These also represent reference states for the purposes of TOTH and TOT(H_2O), noting the latter is omitted from the actual equation state (implicitly resolved). This means for every remaining species (N_R), there is a corresponding equilibrium reaction and hence an equilibrium constant.

The system as shown in Table 2.1 will result in 15 algebraic equations (note water is omitted) and 6 molar contribution balance equations:

$$0 = TOT_J - \sum_{i=1}^{N_S} v_{i,j}\left[S_i\right] \quad i = 1 \ldots N_C \tag{2.21}$$

where $v_{i,j}$ is the stoichiometry entry in the tableau for row i and column j.

TOTH is set to zero (or a constant where the solution background is not known), while the remainder are set from differential equation states. Added to this, there are 10 equilibrium equations, which are a product of the given species and reference species:

$$0 = \{S_i\} - K_i \prod_{j=1}^{N_C} \{S_j\}^{v_{i,j}} \quad i = N_{C+1} \ldots N_S \tag{2.22}$$

Note that equation (2.21) is in concentration units, while equation (2.22) is in activity units. This is discussed further below.

Taking Table 2.1 as an example, the carbonate system ($j = 2$) consists of a single component ($TOT_2 = S_{IC}$), and five species ($i = 2,8,9,12,13$). The molar contribution balance equation ($i = 2$) is

$$0 = S_{IC} - \left[CO_3^{2-}\right] - \left[HCO_3^-\right] - \left[CO_2\right] - \left[NaCO_3^-\right] - \left[NaHCO_3\right] \quad (2.23)$$

The four equilibrium equations to determine the species are ($i = 8,9,12,13$)

$$0 = \left\{HCO_3^-\right\} - 10^{10.33}\left\{H^+\right\}\left\{CO_3^{2-}\right\} \quad (2.24)$$

$$0 = \left\{CO_2\right\} - 10^{16.68}\left\{H^+\right\}\left\{CO_3^{2-}\right\}^2 \quad (2.25)$$

$$0 = \left\{NaCO_3^-\right\} - 10^{1.27}\left\{Na^+\right\}\left\{CO_3^{2-}\right\} \quad (2.26)$$

$$0 = \left\{NaHCO_3\right\} - 10^{10.03}\left\{H^+\right\}\left\{Na^+\right\}\left\{CO_3^{2-}\right\} \quad (2.27)$$

Note that equations (2.23–2.27) are expressed in the form of equations (2.21 and 2.22) (i.e., fully implicit), but could be explicitly expressed. As an example, equation (2.24), explicitly expressed for bicarbonate activity is

$$\left\{HCO_3^-\right\} = 10^{10.33}\left\{H^+\right\}\left\{CO_3^{2-}\right\} \quad (2.28)$$

2.2 NON-IDEALITY/ACTIVITY

Inclusion of calculated activity coefficient corrections allows relaxation of the ideal mixture ($\gamma_i = 1$) assumption by representation of compound activity rather than concentration. In ions, this effectively accounts for ion and solvent interactions that are not accounted for in specific reactions (e.g., ion pairs). For ions, these interactions are a combination of long-range (coulombic) and short-range (molecular dipole) electrostatic interactions (Morel & Hering, 1993). Activity for ionic species is usually lower than concentration that is the 'effective concentration' is reduced. For non-ionic (neutral) species, the activity of the solvent (water) is reduced by the ionic species, which means that its ability to act as solvent to non-ionic species is reduced, and properties such as effective solubility of non-ionic species are therefore decreased. This implies an *increase* in activity for non-ionic species. This also impacts on ionic species, but instead the ionic effects above dominate.

Non-ideality becomes important for functional simulation at a typical domestic wastewater strength or greater (Table 2.2). Specific chemical interactions (ion pairing) need to be included separately as discussed above. None of the IWA models represent non-ideal ion behaviour, though this has been added through the course of this Task Group (Solon *et al.*, 2015). Commercial software generally includes some form of simplified activity correction. A similar level is included in the UCT models (Musvoto *et al.*, 2000a), together with some ion pairing behaviour.

The overall solution properties (in terms of deviation from ideality) are determined by ionic strength (Hamann *et al.*, 2007; Stumm & Morgan, 1996):

$$I = \frac{1}{2}\sum_i z_i^2 \left[S_i^{z\pm}\right] \quad (2.29)$$

Note that S_i is molar concentration, not activity.

Individual activity corrections γ_i are determined by a number of corrections based on ionic strength. Non-ideality has an impact on both ionic and non-ionic species in the liquid phase, as noted above,

Table 2.2 Required non-ideality corrections to achieve pH error of <5%.

Level	Ionic Strength (M)	Wastewater Type	Approach
1	<0.001	Drinking water, clean natural fresh water	No correction required – assume ideal
2	<0.1	Weak industrial WW. All domestic WWTP.	Non-iterative simple correction. Non-ion specific activity coefficient. $K_A' = K_A \gamma_A$ [1]
3	<1 (only ionic species corrected)	Brackish, anaerobic digesters	Full iterative calculation of ion activity using appropriate activity coefficient calculation.
4	<1 (all species corrected)	As above, with gas transfer	As above, with addition of uncharged specie correction (e.g., Table 2.3).
5	<5	Strong industrial, landfill leachate, RO brine	Requires specific ion interaction (SIT) corrections

[1] γ_A is the static activity correction for single valency ions suitable for K_a (generally monovalent for simple corrections).

but simple corrections consider only ionic species. Activity is applicable to chemical reactions and the related equilibria equations, not the charge or proton balance (though it has secondary effects via the implicit nature of equations). The impact of non-ideality on gas liquid transfer and precipitation is substantial via its impact on equilibrium. This influences the activity of free ions in the liquid phase, and thereby determines the chemical/thermodynamic driving force for gas–liquid transfer and precipitation reactions.

Ionic and non-ionic species can be treated separately, with an activity correction for neutral molecules (Table 2.3). There are a number of methods to calculate ionic activity coefficients for different ionic strengths. These can be divided into (1) corrections for ions with general parameters (usually a single set each for monovalent, divalent and trivalent ions), (2) corrections for ionic species which have species-specific parameters, and (3) inclusion of specific species interaction parameters (SIT, Pitzer). All forms include the term z_i^2, which means that multi-valent ions are most strongly impacted, which has a strong effect on precipitation reactions.

Based on Table 2.2, any wastewater modelling problem requires activity correction. Where multi-valent ions are important (e.g., phosphorus precipitation and coagulation), this becomes even more important. However, it may be possible (though risky) to use a simple non-iterative activity calculation based on ionic strength from a previous step. At higher wastewater strengths, it is required to iterate to determine activity (γ_i), as this subsequently influences ionic strength (I) through speciation.

Activity models are presented in Table 2.3, with an increasing complexity and ionic strength range. The basic Debye–Hückel correction breaks down quickly, and has substantial error above $I = 0.1$ M, even for simple solutions (Morel & Hering, 1993).

For common wastewater applications (ionic strength up to 1 M), and based on the review in Table 2.3, the Davies equation is simple, has a relatively broad range, and low-to-moderate errors (± 0.1) for simple solutions (Morel & Hering, 1993) and wastewater applications (Loewenthal *et al.*, 1994). At higher concentrations, specific interactions need to be included for dominant ions, or adjusted parameters need to be used for Davies or better non-SIT models (e.g., WATEQ). A comparison of the different activity models vs measured activities is shown in Figure 2.1.

At very high ionic strengths (>5 M) it is very important to use an appropriate activity correction, and Davies or other simple corrections may actually invert (the second term takes over), and result in an activity correction coefficient of >1 (Thompson Brewster *et al.*, 2017a) (albeit often outside suggested validity ranges of the activity correction model). At an ionic strengths of 1–5 M, such as in landfill leachate, approximately 50% of non-ideality can be explained by activity corrections (level 3), while the other 50% can be explained by specific ion pairing (Tait *et al.*, 2009). Most high-strength activity

Table 2.3 Activity coefficient models.

Level	Type	Model	Ionic Strength	Reference
1	Debye Hückel	$\log(\gamma_i) = -Az_i^2\sqrt{I}$	0–0.05	Hamann *et al.* (2007)
	Günterberg	$\log(\gamma_i) = -Az_i^2\left(\frac{\sqrt{I}}{1+1.4\sqrt{I}}\right)$	0–0.1	Sawyer *et al.* (2003)
	Davies	$\log(\gamma_i) = -Az_i^2\left(\frac{\sqrt{I}}{1+\sqrt{I}} \quad 0.3I\right)$	0–0.7	Morel and Hering (1993)
2	Extended Debye Hückel[1] (PHREEQC)	$\log(\gamma_i) = -Az_i^2\left(\frac{\sqrt{I}}{1+Bc_i\sqrt{I}}\right)$	0–0.5	Langmuir (1997), Parkhurst and Appelo (2013)
	Extended Debye Hückel (WATEQ)[1] (also in PHREEQC)	$\log(\gamma_i) = -Az_i^2\left(\frac{\sqrt{I}}{1+Bc_i\sqrt{I}} - d_iI\right)$	0–1	Merkel *et al.* (2008)
4	SIT	$\log(\gamma_i) = -Az_i^2\left(\frac{\sqrt{I}}{1+B\sqrt{I}}\right) - \sum_j b_{ij}\left[S_j\right]$	>>1 Solubility limit for high solubility salts	Guggenheim and Turgeon (1955)
	Pitzer[2]	Extends SIT for second- and higher-order interactions	>>1 Solubility limit for high solubility salts	Pitzer (1991)
N/A	Non-ionic species	$\log(\gamma_i) = d_iI$ $d_i \sim 0.1$	All	Morel and Hering (1993)

[1]A, B are solution specific. c_i is an ion-specific size parameter d_i, empirical specific to ion i. b_{ij} for interaction of ion i with ion j.
[2]Often developed for single (commonly monovalent) salts to the solubility limit.
$A \sim 0.5$ for water as solvent, or more precisely determined as $A = 1.82 \times 10^6(\varepsilon T)^{-3/2}$ (Stumm & Morgan, 1996) where ε is the solvent dielectric constant (78.3), and T is the solution temperature in K.

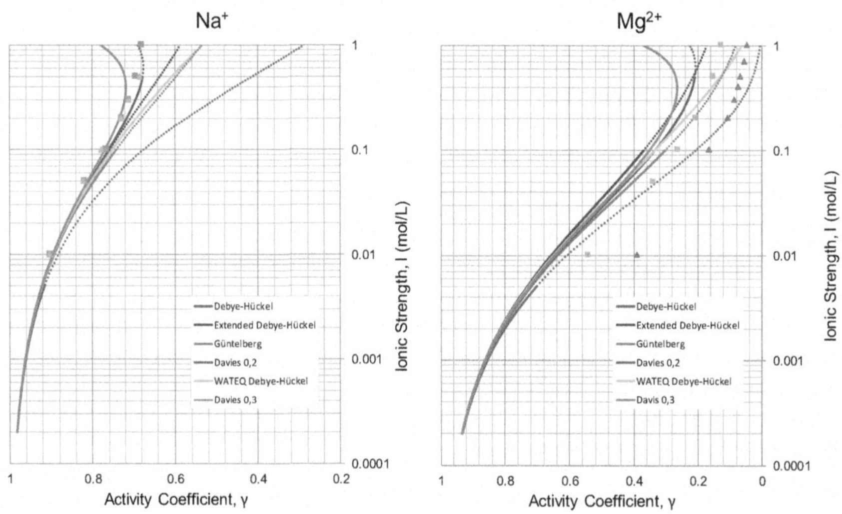

Figure 2.1 Comparison of different activity models for monovalent and divalent ions (Na^+ and Mg^{2+} shown as examples) including data from 1:1 electrolytes from Harned & Owen (1958) (NaOH and NaCl for Na^+), and sulfuric acid (square) and $MgSO_4$ (triangle). Note that measured activity coefficients are net average ($\gamma_\pm$) (updated from Tait *et al.* (2012)).

analysis (particularly Pitzer) has been done against 1:1 (generally monovalent) electrolytic solutions, often using very old data, and more work is needed to develop and validate activity corrections in high-strength wastewater matrixes.

Activity can be included in speciation in two different ways, which are algebraically equivalent:

(a) Activity is used in speciation equilibria (e.g., 2.8). This requires calculation of I, and calculation of γ_i, and/or $\{S_i\}$ or inclusion of γ_i in equilibrium or mass balance equations (see below).
(b) K_a is dynamically (iteratively or not) corrected for γ_i (Loewenthal *et al.*, 1994; Stumm & Morgan 1996). In this case, the general equation:

$$K_A = \frac{\{A^-\}\{H^+\}}{\{HA\}} = \frac{\gamma_A [A^-]_H [H^+]}{\gamma_{HA}[HA]} \tag{2.30}$$

Note that $\gamma_A \equiv \gamma_{A^-}$ and $\gamma_H \equiv \gamma_{H^+}$. The charge is omitted in the above equation to avoid confusion. Equation (2.31) can be reformulated as

$$\frac{K_A \gamma_{HA}}{\gamma_A \gamma_H} = K_A' = \frac{[A^-][H^+]}{[HA]} \tag{2.31}$$

Note that $K_A' = \frac{K_A \gamma_{HA}}{\gamma_A \gamma_H}$ can be further approximated to $K_A'' = K_A/\gamma_A$, since $\gamma \sim 1$ for non-ionic species at lower ionic strengths, and activity of protons can be used instead of concentration ($\gamma_{H^+}[H^+] = \{H^+\}$) in the closure equation (i.e., equation 2.12 or equation 2.13). Note that pH is a measure of proton activity, not concentration. When used iteratively, this represents the simplest form of activity correction. This can be made non-iterative where I is relatively small (i.e., <0.1 M)

Option (a) is better where larger models are used (particularly at higher ionic strengths), or where a generally applicable approach is desired. Option (b) is better for simpler systems, or where a non-iterative approach is taken (e.g., for domestic wastewater only, see Table 2.3). The major reference geochemical models apply option (a) (Allison *et al.*, 1991; Parkhurst & Appelo, 2013).

If option (a) is taken, there are three ways to implement this. The first is explicit calculation of both activity and species concentration (as separate variables), and their use in equations (2.21 and 2.22). This substantially increases the size of the explicit algebraic equation problem. The alternative is reformulation of either equation (2.21) or equation (2.22) as either (for equation (2.21):

(a)

$$0 = TOT_J - \sum_{i=1}^{N_S} v_{i,j}\{S_i\}/\gamma_i \quad i = 1 \ldots N_C \tag{2.32}$$

And leaving equation (2.22) as is

$$0 = \{S_i\} - K_i \prod_{j=1}^{N_C} \{S_j\}^{v_{i,j}} \quad i = N_{C+1} \ldots N_S$$

(b) or leaving equation (2.21) as is

$$0 = TOT_J - \sum_{i=1}^{N_S} v_{i,j}[S_i] \quad i = 1 \ldots N_C$$

and expressing equation (2.22) in terms of molar concentration:

$$0 =_i [S_i] - K_i \prod_{j=1}^{N_C} [S_j] v^{i,j} \quad i = N_{C+1} \ldots N_S \tag{2.33}$$

Either approach leaves both sets of equations consistently either in concentration or activity. The former (equation 2.33) is simpler, since there are multiple species (per component) equations embedded in equation (2.34) (all of which need activity incorporated), whereas there is only one equation per component in equation (2.33) (for all species). Either approach requires recalculation of concentration or activity respectively using calculated activity coefficient. Taking the total inorganic carbon system as an example shown previously in equations (2.23–2.27), only equation (2.23) is changed for the activity approach as

$$\begin{aligned} 0 = S_{IC} &- \{CO_3^{2-}\}/\gamma_{CO3} - \{HCO_3^-\}/\gamma_{HCO3} - \{CO_2\}/\gamma_{CO2} \\ &- \{NaCO_3^-\}/\gamma_{NaCO3} - \{NaHCO_3\}/\gamma_{NaHCO3} \end{aligned} \tag{2.34}$$

Equation (2.35) simplifies substantially in the case that a level 1 activity correction is applied (Debye Hückel, Günterberg, or Davies), since there are common activity corrections for monovalent, divalent, and no activity corrections for uncharged species such that equation (2.35) becomes

$$0 = S_{IC} - \{CO_3^{2-}\}/\gamma_2 - \{HCO_3^-\}/\gamma_1 - [CO_2] - \{NaCO_3^-\}/\gamma_1 - [NaHCO_3] \tag{2.35}$$

where γ_1 is the activity correction for monovalent species, and γ_2 is the activity correction for divalents. Note that with level 1 activity models, concentration equals activity for non-ionic species.

2.3 IMPLEMENTATION AND SOLUTION

The generalised assumption in formulating acid–base and ion pairing speciation equations is that of chemical equilibrium. That is, given that acid–base speciation and ion pairing are extremely fast, the solute-phase reactions can be assumed to be at equilibrium compared with most other processes occurring in wastewater treatment. These equilibrium reactions can be formulated as an algebraic equation set to be solved as is, or it is possible to reformulate the resulting algebraic equations as differential equations, as discussed below.

In the basic algebraic formulation, given the specified TOT_j components from conservation equations, it is possible to solve the algebraic unknowns representing concentrations or activities of the different species (including hydrogen ions from the proton or charge balance).

This algebraic approach results in an implicit set of algebraic equations, which needs an iterative algebraic solver for any substantial level of complexity. While it can be restricted for simplicity (see below) even to a very small number of algebraic unknowns, in most cases, and for generalised application, a numerical solution is required. This results in the wastewater problem being a semi-explicit, index 1, differential and algebraic equation problem (DAE) (Hangos & Cameron, 2001):

$$y' = f(y, z, t) \tag{2.36}$$

$$0 = g(y, z, t) \tag{2.37}$$

where y are differential variables (components and other states), z are algebraic variables (species, ionic strength), and t is the time. Note that the DAE system is index 1 with respect to the liquid-phase reactions system, and that effective boundary conditions still need to be chosen for the larger I/O system to maintain this status.

2.3.1 DAE approach

Formulation and solution as a DAE system is the recommended generalised approach, particularly for complex systems or plant-wide modelling. This separates slow reactions and mass transport as differential equations (equation 2.37) and the equilibrium equation set (equations 2.21, 2.22 + ionic strength/activity as required) as the algebraic equations (equation 2.38), in an index 1 DAE problem. As noted above, an implicit (stiff) solver can solve these problems natively, and this is often the easiest approach. Where other factors (e.g., discontinuities, dynamic inputs etc) make the use of stiff solvers impractical, or for general application, a separate implicit algebraic solver can be used to solve $0 = g(y,z,t)$. This provides the opportunity to optimise the solver for the particular issue of acid–base and ion pairing equilibria, and to optimise the algebraic equation set, in terms of number of variables, calculation speed, and convergence. The reference implementation of the PCM1 utilises nomenclature Z for implicit algebraic variables, which may be activity, ionic strength, or concentration (to separate these from differential state concentration variables S). In this context, S_i, and Z_i are particularly for numerical formulation (compared with $\{S_i\}$ and $[S_i]$, which are generally used in this book for concentration and activity respectively to present the theory).

It is possible to substantially reduce the number of algebraic unknowns by substitution. Simple systems can be made explicit in proton concentration, and even very complex systems can be reduced to one ($[H^+]$) very complicated implicit or even explicit algebraic equation through symbolic manipulation (Takacs, 2008). However, this is not a universally applicable, generalised, nor easily expandable approach. It is often faster to solve multiple simpler implicit equations through several steps than a single very complex implicit (or single step of a very complex explicit equation).

A simple and generalised simplification is to substitute equation (2.22) for each species into equation (2.21) (or its activity form equation 2.33), which reduces $N_S + N_C$ equations to N_C equations (plus ionic strength/activity coeff). This makes even a very complex plant-wide model much more accessible (from 119 to 21 in the case of the PCM1).

Multivariate algebraic equations are generally solved through Jacobian inversions (Newton–Raphson - explicit Jacobian, or Secant - Jacobian approximation). There are rule-based zero finding techniques (e.g., bisection), but most are only applicable to univariate problems (Gough & Priedhorsky, 2017).

2.3.1.1 Algebraic equation implementation details

The proposed method solves the differential equations separately with an explicit ODE solver, and solves the nonlinear algebraic system using an iterative routine, in this case a multi-dimensional version of the Newton–Raphson method.

$$Z_{i+1} = Z_i - J_G(Z_i)^{-1} G(Z_i) \tag{2.38}$$

where Z_i is the vector of equilibrium states $(z_{1,i},\dots,z_{n,i})$ obtained from the previous iteration step i, $G(z_i)$ is a vector containing the values of the set of implicit algebraic equations $(g_1(z_1,\dots,z_n)\dots,g_n(z_1,\dots,z_n))$, which has to be zero in order to satisfy the equilibrium.

The full analytical Jacobian (J_G) is used for calculation of the new state values, which requires symbolic manipulation of the algebraic equations in order to obtain the matrix of all first-order partial derivatives:

$$J = \partial(g_i,\dots,g_n) \Big/ \partial(z_i,\dots,z_n) \tag{2.39}$$

A numerical Jacobian would also be a possibility – i.e., the secant method. This calculates the Jacobian through first-order numerical partial derivative approximations $J = \Delta(g_1,\dots,g_n)_{i,i-1} \Big/ \Delta(z_1,\dots,z_n)_{i,i-1}$, removes the requirement for symbolic manipulation, and results in a more stable convergence, but with more iterations. The iteration is repeated until the error ($\Sigma G(z_i)$) is less than the absolute tolerance, which in our case is set to 10^{-12}. In the reference code, we apply algebraic equation reduction as noted to 21 implicit algebraics (reference code uses both the charge balance used explicitly, as well as a TOTH example), as well as ionic strength and activities. As a result, the implementation can be easily simulated and combined with ASM/ADM/RWQM while ensuring convergence and without significantly reducing simulation speed.

This Newton–Raphson approach works very well from a consistent initial condition (where Z_0 is close to the solution). This is invariably the case when solving from a previous time step, and generally only 1–2 steps are required with very rapid convergence. It is also relatively effective at initial conditions with moderate pH levels, where an arbitrarily chosen Z_0 may be some distance from the solution, with the Newton–Raphson method ensuring rapid convergence. However, where the state vector is poorly defined (e.g., contains large amounts of metallic cations or anions), the resulting pH may be excessively high or low respectively, and the solution can be non-convergent. In these cases, a numerical approach called simulated annealing (SA) (Reeves, 1995) has been used to increase the robustness of the solver and to make the system less dependent on initial conditions (Flores-Alsina *et al.*, 2015). The reference implementation defaults to SA when the gradient (equation 2.40) calculation results in NaN (due to log, root, divisor issues), or the calculated algebraic is negative. The latter implies convergence to numerically correct, but invalid root.

2.3.2 DE approach

As noted above, it is possible to use a differential approach to solving the acid-base and ion pairing system, with equilibria instead formulated as fast differential equations (effectively reformulating z as y and g as f). This may be applied for simplified systems, where there are limitations imposed by the software, where inclusion of algebraics from the equilibrium speciation model results in a higher-order DAE system (note: this is fairly rare), or where there are other numerical interactions with the model algebraic system.

Formulation of the acid–base and ion pairing equations in terms of DEs results in a substantial increase in complexity – for example, 118 additional states per control volume for the plant-wide component list in Section 2.1.3 (dwarfing even large models such as the ADM1 with 30 states). This can be across a large number of control volumes in a plant-wide model. In addition, the acid–base reaction rates are fast (Musvoto *et al.*, 2000b; Stumm & Morgan, 1996), and must be maintained at this level ($>10^8\,d^{-1}$), to avoid model artefacts (errors), particularly in small control volumes, and where diffusion and gas transfer is relevant (e.g., biofilms), as further discussed in Batstone *et al.* (2012).

The mix of fast and slow processes increases model stiffness (large order of magnitude differences in the model state Jacobian $J = \partial f / \partial y$ eigenvalues, representing time constants). This is mainly a problem for forward (explicit) ODE solvers such as ODE45, where the fast process controls the minimum step size (due to numerical solution stability), and the slow process controls the dynamics and hence conversion rate (e.g., towards steady state) (Hangos & Cameron, 2001). Implicit stiff solvers (backward Euler, DASSL, ODE15s, Gear' Stiff, CVODE) are often used to solve such models, but (a) these are inherently slow or unstable when inputs or disturbances are dynamic, or where the model is discontinuous (including derivatives, and (b) implicit solvers are generally very capable at solving index 1 DAE systems, since they include an algebraic solver.

Finally, it is noted that many other model components can induce implicit algebraic loops (e.g., hydraulic recycle streams, feedback controllers), and a DAE approach may be required regardless.

The differential approach has been extensively discussed previously (Batstone *et al.*, 2012; Musvoto *et al.*, 2000b), and a proposed approach for the ADM1 is given in the relevant scientific and technical report (Batstone *et al.*, 2002a). It is most applicable to simplified, single compartment systems, dominated by a small number of key components. An optimal approach where the DE system is used would include pH calculation, and simple non-iterative activity correction through modification of K_i values.

2.4 CONCLUSIONS AND PERSPECTIVE

Larger chemistry models, including non-ideal behaviour appear very complex, and indeed, incorporate a large number of additional implicit algebraic states. It should be noted though that in this chapter almost no contestable parameters have been introduced. That is, all acid–base, and ion pairing equilibrium parameters are fundamental, derived from chemical thermodynamics, and are readily available in reference texts. The highest degree of uncertainty is introduced by the level of activity corrections required in a model and the related parameters.

The acid–base and ion pairing model developed, as well as the related reference code, is generally applicable across the majority of the water and wastewater cycle, and addition of new components (though requiring an understanding of the approach) is a relatively straightforward task, with the addition of a relevant set of equations in the form of equations (2.21 and 2.22). A competent acid–base (and ion pairing) model, incorporating non-ideal behaviour is an essential foundation for all other model components, particularly, but not limited to precipitation, where multivalent ions are disproportionately impacted by non-ideality, and the basic concepts introduced in this chapter are used throughout the rest of this book.

doi: 10.2166/9781780409832_0021

Chapter 3
Precipitation and dissolution

Christian Kazadi Mbamba[1], Paloma Grau[2], Stephan Tait[3], Xavier Flores-Alsina[4], Imre Takács[5] and Damien J. Batstone[1]

[1]*Australian Centre for Water and Environmental Biotechnology, The University of Queensland, St Lucia, Brisbane, QLD 4072, Australia*
[2]*University of Navarra, Manuel Lardizabal 15, 20018 Donostia/San Sebastián*
[3]*Centre for Agricultural Engineering, University of Southern Queensland, Toowoomba, 4350, Australia*
[4]*Department of Chemical and Biochemical Engineering, PROSYS Research Center, Technical University of Denmark, Building 229, DK-2800, Kgs. Lyngby, Denmark*
[5]*Dynamita, Roquesteron, Provence-Alpes-Côte d'Azur, France*

3.1 THEORY

Precipitation refers to the formation of a solid from solutes in a supersaturated solution under certain conditions (Myerson, 1993). There are a range of precipitation processes, including crystallisation, ion complexing, and ion-complex adsorption. Particularly, the latter may result in variable reaction stoichiometry, and is not effectively represented by a fixed stoichiometry approach as in the first part of this chapter. An alternative approach for hydroxy-ion and aluminium phosphorus precipitation is discussed further in Section 3.22. The process of precipitation itself may have multiple stages as further discussed below.

Reactants for precipitation are generally ionic solutes, but non-ionic compounds can also form precipitates, commonly through dehydration, for example,

$$SiO_{2(s)} + 2H_2O \leftrightarrow H_4SiO_4 \tag{3.1}$$

Precipitation is widespread in the water cycle, with common precipitating cations being calcium (as Ca^{2+}), magnesium (as Mg^{2+}), aluminium (as Al^{3+}), iron (as Fe^{2+} and Fe^{3+}), as well as ammonium (NH_4^+) being involved in multi-cation minerals such as struvite ($NH_4MgPO_4{\cdot}6H_2O$). Common precipitating anions are carbonates (CO_3^{2-}), phosphates (PO_4^{3-}), sulfates (SO_4^{2-}), and sulfides (S^{2-}), and hydroxides (OH^-) particularly for multi-anion precipitates or at high pH (e.g., $Ca(OH)_2$). Precipitates are commonly chemically hydrated (note struvite above) meaning that water forms a part of the precipitate chemical structure.

Precipitates may be amorphous or crystalline. Amorphous precipitates lack regular crystalline structure, generally form rapidly, and have relatively low stability, meaning they will dissolve to their respective solutes rapidly under appropriate conditions. Crystalline precipitates have a crystal structure and often pronounced crystal faces, form more slowly, and generally have a higher degree of stability and comparatively lower solubility. Minerals with the same anions may form both amorphous and crystalline precipitates, with the amorphous form commonly transiting to the more stable crystalline form over time.

3.1.1 Solubility and equilibrium in precipitation

Precipitation is governed by aquatic chemistry equilibrium, as outlined in Chapter 2. Precipitation/dissolution is a reversible process, with redissolution being the reverse of precipitation. The common protocol for expressing the equilibrium for precipitation/dissolution reactions is with solutes as products and precipitates as reactants as

$$A_aB_{b(s)} \leftrightarrow aA^{z+}{}_{(aq)} + bB^{y-}{}_{(aq)} \tag{3.2}$$

Note that for the reaction to balance, $a{\cdot}z = b{\cdot}y$, to strictly form a neutral-charged mineral phase. Since the precipitate $A_aB_{b(s)}$ is in a separate essentially pure phase, it does not influence the equilibrium, because the activity of pure phases is usually taken to be unity. Note that in Chapter 2, water as the solvent also does not appear in the equilibrium equation. Hence, the general equilibrium equation (equation (2.3)) becomes the following for precipitation reactions:

$$K_{sp} = \left\{A^{z+}\right\}^a \left\{B^{y-}\right\}^b \tag{3.3}$$

Here, the equilibrium constant K_{sp} is commonly referred to as the solubility product constant, because equation (3.2) is written in the direction of the dissolution reaction. For convenience (because the numbers are commonly very small), a log scale is commonly used, with $pK_{sp} = -\log(K_{sp})$, where pK_{sp} is provided in reference texts. Note that the equilibrium relationship is written in terms of activity, not concentration.

Solubility refers to the maximum amount of precipitate that will dissolve in water solution at equilibrium. This is less rigorous than the definition of solubility by the solubility product, and is only relatable by calculation, as discussed in the following example.

Calcium carbonate (calcite) forms (or dissolves) according to the following reaction:

$$CaCO_{3(s)} \leftrightarrow Ca^{2+}{}_{(aq)} + CO_3{}^{2-}{}_{(aq)} (K_{sp} = 3.3\times10^{-9};\, pK_{sp} = 8.48) \tag{3.4}$$

The related equilibrium equation is

$$K_{sp} = 4.3\times10^{-9} = \left\{Ca^{2+}\right\}\left\{CO_3{}^{2-}\right\} \tag{3.5}$$

Precipitation will only occur when the concentration of solutes exceeds the threshold solubility product. Calcium carbonate has solubility of 0.013 g L^{-1} (0.00013 M). That is 0.013 g of $CaCO_3$ will dissolve into 1 L of pure water. One might note that K_{sp} is not simply $(0.00013)^2$. This is because both the Ca^{2+} and $CO_3{}^{2-}$ form multiple ion pair species in the aqueous phase, including $CaOH^-$, $HCO_3{}^-$, and $CO_{2(aq)}$, and even $CaCO_{3(aq)}$, as described in Chapter 2. This decouples apparent solubility (the amount of mineral which dissolves), from solubility product constant (the precipitation driving force).

The form of equation (3.5) also introduces a very important concept influencing precipitation, namely, the common-ion effect; this dictates that the equilibrium can be shifted to cause precipitation and reduce the concentration of one participating ion by adding more of the counter participating ion. For example, in the above reaction, $CO_3{}^{2-}$ anions can be added as a different salt Na_2CO_3 to the water solution directly after dissolving the 0.013 g L^{-1} of $CaCO_3$. Now, to attain equilibrium and thereby satisfy the K_{sp} value, $CaCO_3$ precipitation would need to occur to shift the reaction in equation 3.4 towards the left. This would reduce the total dissolved calcium concentration, so that it is now less

Table 3.1 Different expressions used for supersaturation ratio and SI.

$\Omega = \dfrac{IAP}{K_{sp}}$	$\Omega = \left(\dfrac{IAP}{K_{sp}}\right)^n$ n is commonly $1/\upsilon$ where υ is the number of ions participating in the reaction	$SI = \log\left(\dfrac{IAP}{K_{sp}}\right)$
Bouropoulos and Koutsoukos (2000)	Jones (2002), Mullin (1992)	Ariyanto *et al.* (2014), Bhuiyan *et al.* (2008), Galbraith and Schneider (2014), Wang *et al.* (2006)

than the stoichiometry in equation (3.4). This example demonstrates a practical application of the common ion effect, whereby calcium hardness in water has been reduced by inducing precipitation of $CaCO_3$ by adding carbonate in the form of soda ash.

Precipitation/dissolution occurs relatively slowly in comparison with acid base and ion pairing reactions. This means that within practical wastewater treatment time scales, a water solution can be in a state other than solubility conditions, which could induce precipitation or dissolution to approach equilibrium. To identify whether precipitation (or dissolution) conditions exist, it is necessary to define the current system state using an ion activity product (IAP). This is defined as

$$IAP = \left\{A^{z+}\right\}^c \left\{B^{y-}\right\}^d \tag{3.6}$$

Note that equation (3.6) defines the current state, while equation (3.3) defines the precipitation equilibrium state. When $IAP > K_{SP}$, thermodynamics is driving towards precipitation and the aqueous liquid phase is said to be oversaturated, while when $IAP < K_{SP}$, thermodynamics is driving towards dissolution and the aqueous liquid phase is said to be undersaturated.

The saturation extent of an aqueous system can also be expressed as the saturation ratio or oversaturation, Ω, expressed as follows:

$$\Omega = \left(\frac{IAP}{K_{sp}}\right)^{1/\upsilon} \tag{3.7}$$

where υ is the number of ions in the molecular formula of the solid phase (e.g. $\upsilon = 2$ for $CaCO_3$).

The ratio (Ω) of IAP and K_{SP} represents the driving force for precipitation or redissolution (generally called saturation extent), as shown below. A mathematical variant of oversaturation ratio is the saturation index or SI value, expressed as per Table 3.1. Supersaturation ratio (Table 3.1) is commonly included in rate expressions during modelling as further described later in this chapter, including for crystal formation, growth, and/or dissolution.

3.1.2 Mechanisms of mineral formation

In the precipitation process, three steps can be distinguished: nucleation (crystal birth), crystal growth and ageing (Figure 3.1).

3.1.2.1 Nucleation

Precipitation or formation of a mineral particle from a solution starts with a nucleation process, considered as the formation of new mineral solid phase. This solid phase then provides the surface area for further precipitate to deposit on, causing the precipitating particles to grow. Nucleation has a high associated activation energy, and thereby requires a sufficient level of supersaturation as a driving force to overcome the phase energy barrier and induce nucleation. Nucleation can be divided into primary and secondary nucleation.

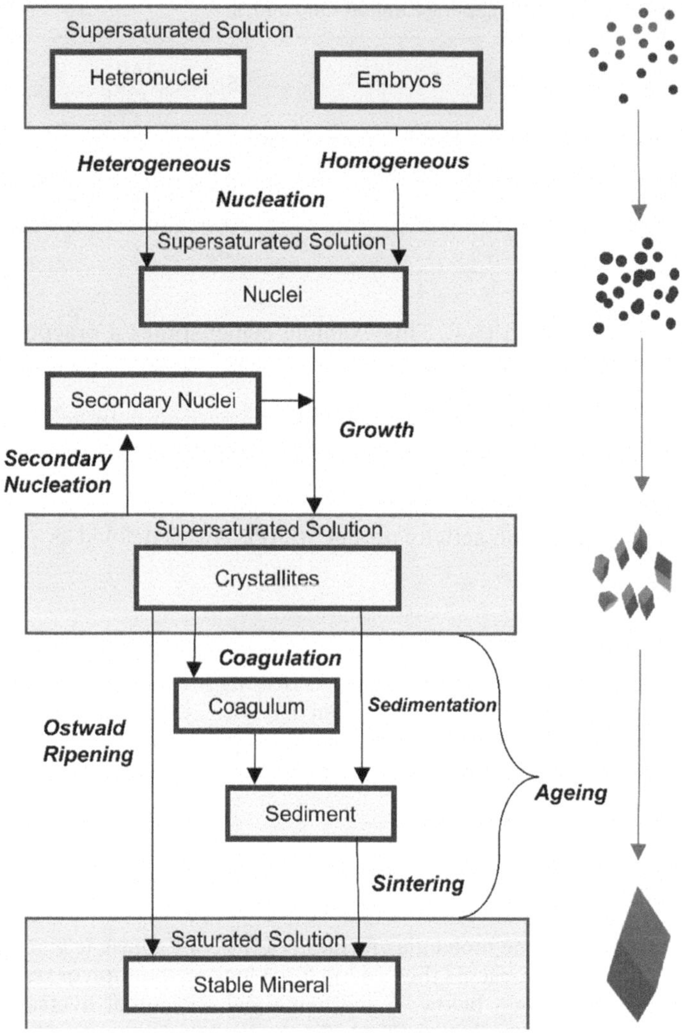

Figure 3.1 Stages of precipitation processes. (*Source*: Modified from Le Corre *et al.*, 2009 and Nielsen, 1970).

Primary nucleation can either be homogenous or heterogenous. Homogenous nucleation can occur when nuclei form spontaneous embryos, or homogenous nuclei from a highly supersaturated solution in the absence of any foreign particles or surfaces. By contrast, heterogenous nucleation takes place on non-mineral phase solid surfaces, whereby the presence of the foreign particles induces heterogenous nucleation by lowering the activation energy barrier. Consequently, primary heterogenous nucleation can be induced at a much lower supersaturation than primary homogenous nucleation. For example, during wastewater treatment, inadvertent precipitation is likely to be heterogenous nucleation owing to the high concentrations of impurity particles (i.e. suspended solids and organic matter) in wastewater.

Secondary nucleation is distinct from primary nucleation, instead occurring in the presence of particles of the precipitating mineral phase, with these particles having a strong catalysing effect and considerably reducing the activation energy barrier for nucleation to occur. As a result, secondary

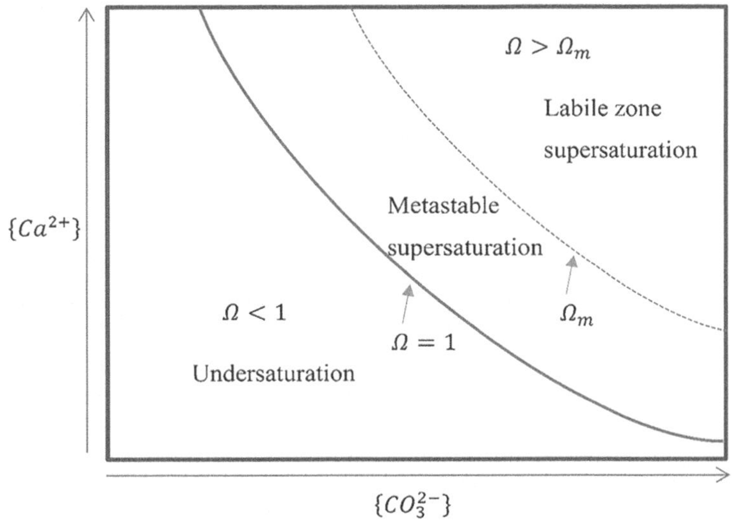

Figure 3.2 Phase diagram showing states of a solution during the precipitation process.

nucleation can occur at much lower levels of supersaturation than either heterogenous or homogenous nucleation (Mullin, 2001).

3.1.2.2 Metastability

The precipitation reaction rate (nucleation or growth) is enhanced, or even enabled by the presence of existing crystal or other solid catalysts (foreign particulate impurities). This also influences the saturation extent required to induce nucleation under different conditions (e.g. presence or absence of foreign particles or mineral phase). This means that there is a metastable zone where IAP $> K_{SP}$ but spontaneous nucleation may not occur because the saturation extent is inadequate. This is explained using the concept of metastability as shown in Figure 3.2. Considering this, there are three possible zones.

- Undersaturated (in the undersaturated zone), meaning that mineral phase can dissolve if present or when added to the solution (IAP $< K_{SP}$ or $\Omega < 1$). Precipitation is thus impossible.
- Metastable (in the metastable zone situated between the solubility curve ($\Omega = 1$) and the nucleation threshold (Ω_m, limit of metastability), meaning that the solution is saturated (IAP $> K_{SP}$) but spontaneous nucleation cannot occur because of inadequate saturation extent. The growth of existing mineral phase can however be induced by seed addition.
- Oversaturated (in the labile or unstable zone), with solute concentration substantially exceeding the equilibrium value (IAP $>> K_{SP}$ or $\Omega > \Omega_m$) causing spontaneous nucleation. Precipitation can be rapid and abundant. This only applies to homogeneous nucleation, which is rarely relevant for wastewater systems.

The position of the nucleation threshold depends on nucleation type and is usually at a lower saturation extent for secondary nucleation than for primary heterogenous nucleation, which in turn is at a lower supersaturation extent than for primary homogenous nucleation.

3.1.2.3 Particle growth

Mineral particle growth follows the nucleation process, where mineral embryos or larger mineral particles grow in size as long as the solution is supersaturated. Mineral particle growth is often described as a two-step process: transport of the solute to the particle surface and integration from liquid-to-solid phase on the particle surface (Myerson, 1993). Mass transport of ions from the supersaturated solution to the mineral surface is dictated by diffusion. Following transport to the mineral surface, adsorption and reaction into the mineral surface occurs. Thus, the overall, rate of growth can be controlled either by mass transport or by surface reaction (Brečević & Kralj, 2007; Nielsen, 1970). Mineral precipitate crystals such as struvite tend to grow lengthwise, rather than laterally, which means that a growth-dominated process results in longer (needle) crystals. At low supersaturation, crystal growth is faster than nucleation, resulting in larger crystal-size distribution. By contrast at higher supersaturation, crystal nucleation prevails crystal growth, ultimately resulting in smaller crystals with flat configuration (Abbona & Boistelle, 1979). The shape of crystal minerals is dictated by the growth rate of respective crystal faces, with fast-growing faces generally disappearing and slow-growing faces being pronounced in the final crystal shape.

Agglomeration is a secondary particle formation process and can become the main crystal growth mechanism in systems with high levels of supersaturation (Jones, 2002). Agglomeration is chiefly responsible for formation when the smaller particles (crystallites), rather than ionic species or molecules collide and cluster together while in suspension via electrostatic formation and cementing, leading to rapid particle-size enlargement.

3.1.2.4 Ageing

Ageing refers to irreversible changes (e.g., decreased interfacial area) that take place in a precipitate after its formation (Mullin, 2001), involving particle consolidation and progressive dissolution and transition to more stable mineral phases. Ageing takes place through ripening, phase transformation and crystallisation of amorphous precipitate. Ostwald ripening refers to process of particle coarsening for a precipitate that remains in contact with its mother liquor for a very long time. During this process, smaller particles (of higher solubility) dispersed in their own saturated solution undergo dissolution to allow the solute to be deposited subsequently onto the larger particles (Nielsen, 1970). Another frequent occurrence of ageing takes place when the precipitation of a metastable phase is followed by a phase transformation into the most stable mineral phase. This agrees with Ostwald's rule of stages, in which a crystallisation process starts with a less stable (amorphous) phase, then ends with a crystalline phase as discussed above. The metastable phase may be an amorphous precipitate, a polymorph of the finale hydrate material, a hydrated species or some system-contaminated substance (Mullin, 2001).

To illustrate phase transformation, precipitation of some minerals, in particular calcium carbonate and calcium phosphate, can take place in a variety of different mineral forms (crystalline vs. amorphous, and/or various crystalline polymorphs) (Barat *et al.*, 2011; Plummer & Busenberg 1982; van Langerak *et al.*, 1999). For example, depending on the pH, supersaturation, and the chemical composition of the solution, three anhydrous crystallite polymorphs of calcium carbonate may precipitate out of solution. These are in order of increasing solubility: calcite, aragonite and vaterite. Calcite is the most thermodynamically stable phase at ambient temperature and pressure, whereas aragonite is an unstable amorphous precipitate, which nucleates first at relatively high supersaturation, then dissolves when the solution becomes undersaturated with respect to aragonite and then follows on with the precipitation of more stable less-soluble forms such as calcite. Calcite is the form of calcium carbonate which will ultimately prevail because it is the most stable calcium carbonate mineral with the lowest solubility. Less stable forms such as aragonite would therefore undergo subsequent transformation into crystalline phases (calcite) (Brečević & Nielsen, 1989).

Finally, ageing takes place when amorphous precipitates undergo transformation into a crystalline product. Hydrous oxides such as ferric hydrous oxides (HFO) are examples of amorphous precipitates that can transform into a crystalline precipitate on ageing. The freshly formed HFO are highly reactive and provides a number of adsorption sites for ions on its surface. Through the ageing process, the molecules of HFO lose reactivity and develop a more compact structure and less accessible sites. This amorphous precipitate is important for chemical phosphorus removal with metal salts (ferric and alum) in wastewater treatment. The mechanism is complex and takes place through two fundamental steps including adsorption of phosphate on HFO and co-precipitation of phosphate into the HFO structure (Smith *et al.*, 2008; Szabó *et al.*, 2008). Adsorption of phosphate species onto HFO leads to production of HFO flocs with bound phosphate and this depends on its structure and reactivity, which is reduced over time as a result of ageing. The mechanism of HFO ageing and phosphate complexation explains precipitates, containing variable ratios of metals that form during chemical phosphorus removal (Szabó *et al.*, 2008).

3.1.3 Factors affecting precipitation rate, equilibrium, and precipitate morphology

Precipitation is a multi-mechanism complex process, with rate, and morphology tightly linked. A high degree of supersaturation results in high driving forces, and high rates enhance primary and secondary nucleation, resulting in a rapid precipitation rate, but with a small crystal size. For mineral precipitates, high rates also result in a higher length aspect (i.e., needles), since agglomeration is minimised, and growth is increased on the longitudinal surface (where activation energy is minimised) – (Durrant *et al.*, 1999; Stumm & Morgan, 1996). This is particularly important where precipitation is used to generate specific products (e.g., to achieve a specific product for fertiliser use), or where subsequent separation requires a large particle size (Le Corre *et al.*, 2009). As indicated above, slower crystallisation rates (induced by a lower SI, often with a suitable seed to overcome nucleation) allow agglomeration to dominate, and hence result in lower aspect, and larger crystals.

Saturation extent. The primary factor which influences whether precipitation occurs, and its rate, is supersaturation extent, quantified as SI or similar (Table 3.1) and several other factors (e.g., pH, temperature, wastewater composition) influence precipitation via affecting saturation extent. High saturation extents promote nucleation, which results in a large number of small crystals (formed rapidly), while low saturation extents promote growth over nucleation, and can result in large crystals (formed slowly).

pH. pH has a substantial impact, mainly by controlling SI, controlling availability of both cationic and anionic solutes. At high pH, the anionic partner (e.g., CO_3^{2-}, PO_4^{3+}) has an increased relative concentration, while at low pH, the cationic partner (e.g., Ca^{2+}, Mg^{2+}) has an increased concentration. There is therefore an optimum pH which maximises SI for a given concentration of solutes, which depends on the pK_a of the anion and cation. The limiting solute (anionic or cationic) is defined by that which, for an equimolar solution, has a lower concentration/activity at a neutral pH. For calcium and magnesium carbonates and phosphates, the anion is the controlling solute, and precipitation is generally enhanced at high pH values. However, for other compounds, low pH may limit precipitation, particularly where the precipitating anion is a metal hydroxide forming an amorphous precipitate with phosphate (e.g., $AlOH^{x+}$, $FeOH^{x+}$) – see later in this chapter.

Mixing and liquid shear. Increased mixing enhances transport of solute to the crystal surface, and high shear rates cause increased nucleation, through both a decrease in nucleation activation energy, and breakage of larger crystals (Durrant *et al.*, 1999). Mixing also enhances agglomeration through particle contact, and reduces length aspect through mechanical shear, particularly for struvite minerals (Wang *et al.*, 2006). Therefore, high shear increases precipitation rate, but can reduce average precipitate size (Elduayen-Echave *et al.*, 2020). The effect is expected to be highly non-linear, and there is generally an optimum where rate is maximised, and particle shape and size is optimised. Overall, liquid shear is an important control parameter with independent mechanisms to saturation extent (Le Corre *et al.*, 2009; Ohlinger *et al.*, 1999; Wang *et al.*, 2006).

Temperature. Temperature has a direct impact on precipitation through changes in the K_{SP} as described by the van't Hoff equation (equation (2.5)). Most solubilisation reactions are exothermic (shown in the form and direction of equation (3.2)). Therefore, solubility tends to increase with increasing temperature and K_{SP} decreases or pK_{SP} increases, but the effect is generally small, and for some minerals (e.g., urea) the solubilisation reaction is endothermic.

Temperature can also strongly influence crystal growth rates, since it has a considerable impact on mechanisms of particle growth such as relative diffusion rates of diffusion and surface integration steps. In this respect, high temperatures of crystallisation lead to diffusion-controlled growth, while low temperatures lead to surface integration controlled growth (Jones, 2002). Moreover, the rate of precipitation often increases at a high temperature, whereas crystal size, shape and type can be affected by the temperature. As an illustration, a crystal habit (shape or morphology) modification was observed between struvite crystallised at 25°C compared to those obtained at 37°C. At 25°C, 'struvite crystals were rectangular and prismatic, whereas at 37°C they were found square and thick' (Boistelle *et al.*, 1983).

Impurities. The growth rate of minerals, nucleation and morphology in a metastable solution can be influenced by the presence of foreign particles or impurities. For example, calcite formation can be inhibited in the presence of foreign ions such as magnesium (Mg^{2+}) or phosphate (PO_4^{2-}). These impurities may block the active sites where participating ions would have integrated into the growing mineral surface (Myerson, 1993). Impurities have a significant effect on crystal growth rates (typically decreasing) and can also change the effect of supersaturation on crystal growth rates. However, the influence of impurities has been shown to be much weaker than that of supersaturation, mineral phase concentration and temperature (Kazadi Mbamba *et al.*, 2015a). Accordingly, the task group recommends that correction for impurity impacts using dedicated inhibition-style terms (van Langerak *et al.*, 1999) be considered optional depending on the level of model application.

3.2 PRECIPITATION IN WASTEWATER TREATMENT

In wastewater treatment, a large number of potential precipitates can form, with precipitate types changing depending on the process and conditions. Key factors are pH changes and availability of solutes due to biological and chemical reactions, commonly related to change in the oxidation–reduction regime.

Key precipitating species in wastewaters include carbonate and phosphate minerals of calcium and magnesium (Barat *et al.*, 2011; Musvoto *et al.*, 2000a; van Rensburg *et al.*, 2003), magnesium ammonium phosphate (struvite), sulfide minerals (at anaerobic conditions) and aluminium and iron phosphates, hydroxides, and sulfides (Batstone *et al.*, 2012). Locations where specific precipitates commonly occur are shown in Figure 3.3, and a summary of reactions and pK_{SP} is provided in Table 3.2.

In line with the Oswald rule of stages, some precipitates are also capable of forming a variety of different minerals, including amorphous and crystalline polymorphs (Brečević & Nielsen 1989; Kralj *et al.*, 1997; Meyer & Weatherall 1982). This is consistent with the Ostwald rule of stages as above.

Two broad systems dominate; the calcium–magnesium–carbonate–phosphate system, with moderate solubilities, and the iron–aluminium–phosphate sulfide system, with low solubilities. Both of these systems have a large number of potential precipitates, with multiple cations and anions (in competition), and both systems are relevant throughout the wastewater treatment plant, and under a range of optimal pH conditions.

Most research focuses on one of these systems, either in a single unit or across the treatment system, with combined approaches relatively rare in the scientific literature (though not uncommon in modelling software).

A summary of precipitate chemistry and equilibrium is shown in Table 3.2. This shows chemical formulae of common precipitates, the stoichiometry in reaction matrix format, approximate formation rates in a wastewater environment, precipitation constants and, for reactants that are commonly limiting, the concentration of the limiting reagent set.

3.2.1 Calcium and magnesium carbonates and phosphates

This system is relevant plant wide, and multiple groups have conducted studies on the system, including the University of Cape Town (UCT) group (Musvoto *et al.*, 2000b; van Rensburg *et al.*, 2003), and the University of Queensland group (Kazadi Mbamba *et al.*, 2015b). The focus has been mainly on

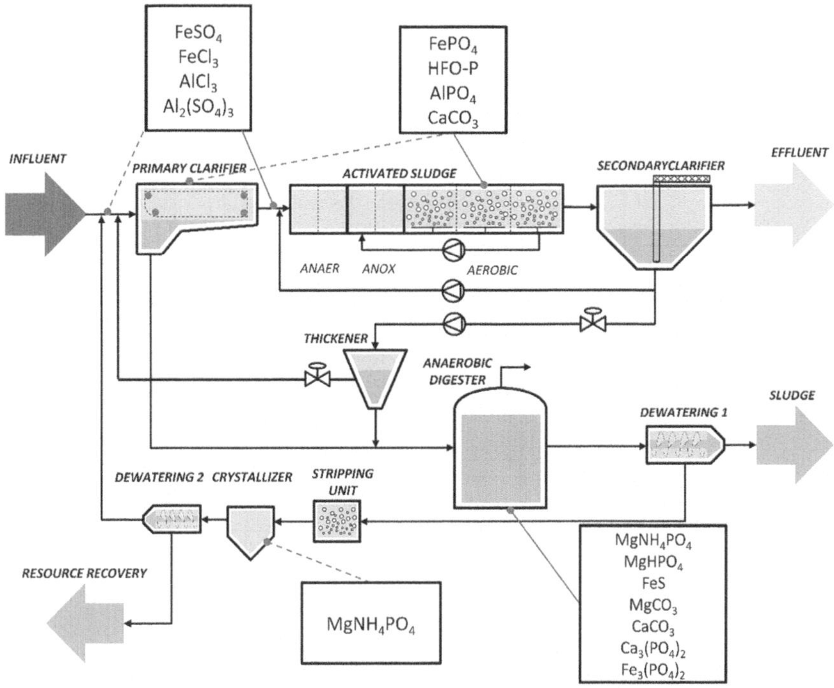

Figure 3.3 Location of chemical addition for phosphorus removal and common precipitates.

anaerobic and aerobic digesters, and digester supernatants, but the effect has also been considered in a plant-wide context.

A large number of precipitates form from these cations and anions, as shown in Table 3.2. At high phosphate concentrations, Ca^{2+} and Mg^{2+} will selectively bind with phosphate to form struvite and amorphous and crystalline calcium and magnesium phosphates, with, phosphate availability normally limiting precipitation extent. At low phosphate concentrations, and high Ca/Mg concentrations, amorphous and crystalline calcium and magnesium carbonates will form.

Some compounds are extremely stable, but the activation energy for formation is so high (i.e., kinetics are too low, or nucleation precursor required) that they do not form practically in a wastewater environment. Examples include ordered dolomite (Banerjee, 2016) and hydroxyapatite (Mohd Pu'ad *et al.*, 2020). Formation of faster-forming competitive precipitates (e.g., amorphous calcium phosphate) limit the concentrations to below nucleation thresholds, and subsequent transformation to stable minerals such as hydroxyapatite is extremely slow, or requires special conditions (Pan *et al.*, 2010).

Apart from struvite, almost all minerals are polymorphic. At moderate–high pH levels (7.5+), where there is sufficient ammonia, struvite will dominate over metal phosphates (as shown in Table 3.2) (Kazadi Mbamba *et al.*, 2015b; van Rensburg *et al.*, 2003). It forms faster, and has a lower residual phosphate concentration than calcium and magnesium phosphate (Barat *et al.*, 2011). Brushite dominates at moderate to sub-neutral pH levels (<7.5), and moderate–low pH brushite precipitation by calcium addition is a viable form of phosphorous recovery (Shaddel *et al.*, 2019).

3.2.2 Iron and aluminium phosphates and sulfides

The iron–aluminium–phosphorus–sulfur system interacts, particularly throughout wastewater treatment, but also in natural systems depending on elemental availability and oxidation state. In

Table 3.2 Summary of relevant Mg/Ca precipitates and precipitation equilibria constants.

	Formula	Formation Rate	Ca^{2+}	Mg^{2+}	NH_4^+	H^+	PO_4^{3-}	CO_3^{2-}	OH^-	pK_{sp}	Limiting Conc (M) $\log_{10}$ (1)
Phosphates											PO_4 (net)
Struvite	$MgNH_4PO_4 \cdot 6H_2O$	Fast		1	1		1			12.6	−5
Newberyite	$MgHPO_4 \cdot 3H_2O$	Fast		1		1	1			5.8	8
Bobierrite	$Mg_3(PO_4)_2$	Slow		3			2			25.2	−5
Trimagnesium phosphate	$Mg_3(PO_4)_2 \cdot 2H_2O$	Moderate		3			2			23.1	−4
Brushite (dicalcium phosphate)	$CaHPO_4 \cdot 2H_2O$	Fast	1			1	1			6.6	−5
Hydroxyapatite	$Ca_{10}(PO_4)_6(OH)_2$	Negligible	10				6		2	114	−9
Amorphous calcium phosphate	$Ca_3(PO_4)_2 \cdot XH_2O$	Moderate	3				2			25.4	2
Carbonates/hydroxides											Ca/Mg
Magnesite	$MgCO_3$	Moderate		1				1		7.46	−3
Nesquehonite	$MgCO_3 \cdot 3H_2O$	Moderate			1			1		5.19	−1
Calcite; Aragonite	$CaCO_3$	Moderate	1					1		8.34, 8.48	−4
Dolomite (disordered)	$CaMg(CO_3)_2$	Slow	1	1				2		16.54	−4
Dolomite (ordered)	$CaMg(CO_3)_2$	Very slow	1	1				2		17.09	−4
Calcium hydroxide	$Ca(OH)_2$	Fast	1						2	5.2	4
Brucite	$MgOH_2$	Fast		1					2	11.16	1

*Data and stoichiometry from Stumm and Morgan (1996). Additional information available in Musvoto *et al.* (2000b) (1) Limiting concentration, that is, the concentration above which precipitation starts to occur for net PO_4 ($[S_{IP}]$) and metal ions (Mg^{2+}/Ca^{2+}) at pH 7, counter ions 0.1M for $Ca^{2+}/Mg^{2+}/NH_4^+$, and 0.1 M for ($[S_{IC}]$).

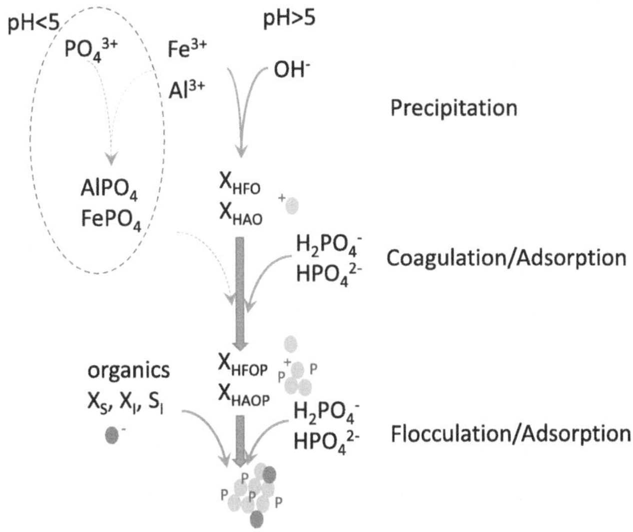

Figure 3.4 Phosphorus reactions with iron and aluminium. (*Source*: Modified from Batstone *et al.*, 2018).

the mainline, under aerobic conditions, the cations of iron (Fe^{3+}), and aluminium (Al^{3+}) (or rather, their hydroxy complexes – see below) react with phosphate to form a range of metal phosphates and hydroxyphosphates (Batstone *et al.*, 2018) with low residual solubility. This is used deliberately to remove phosphates in chemical phosphorus removal. Particularly iron also complexes with sulfide. Under aerobic conditions, reduced and elemental sulfur is oxidised to sulfates. Iron and aluminium sulfates have high solubility.

While ferric and aluminium will form balanced metal precipitates (at longer time scales and low pH), under most practical conditions in wastewater treatment plants, the phosphorus removal process is dominated by phosphate adsorption on aluminium (HAO) and ferric hydroxide (HFO) (Figure 3.4). Hydroxide ions react with iron and aluminium to form iron and aluminium hydroxide (Ferguson & King, 1977; Hauduc *et al.*, 2015; Smith *et al.*, 2008) (Figure 3.4). Hydroxide will outcompete phosphate for the cations under most conditions, and above pH 5, no ferric phosphate is formed (Smith *et al.*, 2008). Iron and aluminium hydroxides as a flocculant precipitate then adsorb phosphate ions as $H_2PO_4^-$, HPO_4^{2-} and PO_4^{3-} (generally the former two), as well as other compatible anions, including organics. The capacity to remove phosphate is governed by the availability of adsorption sites on the metal hydroxide (Hauduc *et al.*, 2015; Smith *et al.*, 2008). Ferric and aluminium have very similar equilibrium, with a slightly lower pH optimum (for P removal) for ferric (Recht and Ghassemi, 1970; Smith *et al.* 2008) (Table 3.3).

In anaerobic conditions, sulfate is reduced to sulfide (Hao *et al.*, 2014), and iron is reduced from ferric (Fe^{3+}) to ferrous (Fe^{2+}) (Wang *et al.*, 2019). Sulfide may induce phosphate release from iron phosphates, which has strong implications when it comes to phosphorus recovery (Wilfert *et al.*, 2020). Ferric hydroxide has a far higher solubility than ferrous hydroxide, and Fe^{2+} (which is generated from Fe^{3+} under reducing conditions) also forms Vivianite ($Fe_3(PO_4)_2{\cdot}8H_2O$ (Wilfert *et al.*, 2016). While aluminium does not precipitate with sulfide, both forms of iron precipitate strongly with sulfide (Gutierrez *et al.*, 2010).

Therefore, the system forms a complex set of interactions, with iron and aluminium forming amorphous hydroxy phosphates of variable stoichiometry under aerobic conditions which is utilised for mainline phosphate removal. These precipitates then pass through to the sludge line, where

Table 3.3 Iron and aluminium precipitates.

Name	Formula	pKsp	Limiting Fe/Al Conc (M) $\log_{10}^{(1)}$	Condition
Ferric phosphate (crystalline)	$FePO_4$	26.4	−16.1	Ae, low pH
Ferric phosphate (amorphous)	$FePO_4$	21.8	−11.5	Ae, low pH
HFOP (ferric)	$Fe_x(OH)_y(PO_4)_z$	Varies	N/A	Ae
Ferric hydroxide	$FeOH_3$	37.4	−16.4	Ae
Ferrous hydroxide	$FeOH_2$	15.1	−1.1	AnAe
Ferrous sulfide	FeS	19.8	−8.3	AnAe
Vivianite (ferrous phosphate)	$Fe_3(PO_4)_2 \cdot 8H_2O$	35.8	−12.8	AnAe
Aluminium phosphate	$AlPO_4$	21	−10.7	Ae/Anae, low pH
Aluminium hydroxide	$Al(OH)_3$	32.9	−11.9	Ae/Anae
HAOP (aluminium)	$Al_x(OH)_y(PO_4)_z$	varies	N/A	Ae/Anae

Values from Ferguson and King (1977) and Smith *et al.* (2008) (1) Indicates Fe/Al concentration above which precipitation can occur (M, $\log_{10}$ scale. Ae = aerobic, AnAe = anaerobic, Ae/Anae = anaerobic or aerobic.

anaerobic conditions reduce sulfates to sulfides, and ferric iron (Fe^{3+}) to ferrous (Fe^{2+}). The phosphate and sulfide compete for Fe^{2+} to form vivianite and iron sulfide. Polysulfide can bind with iron to form pyrite, which occurs due to incomplete reduction of sulfate to sulfide.

3.2.3 Others (sulfates, silicates, heavy metals)

Most sulfates are highly soluble, except calcium sulfate with a $pK_{SP} = 4.6$ (Tait *et al.*, 2009), with even calcium sulfate precipitating in the >10 g Ca/L (Tait *et al.*, 2009), which is relevant for industrial, but not domestic wastewaters. It is noted that ion pairing and ion activity has a substantial impact on effective calcium sulfate solubility, and this highlights the importance of including these in a generalised model (Tait *et al.*, 2009). Sulfate will also precipitate with barium ($pK_{SP} = 10$) (Runtti *et al.*, 2016), and lead ($pK_{SP} = 8$) (but relevant concentrations are not generally seen in wastewater).

Heavy metals considered include Co, Ni, Cu, Zn, Cd, Hg, Sn, and Pb. These have varying affinity for (in order of most soluble to least soluble) CO_3^{2-}, PO_4^{3-}, and S^{2-} (Generalic, 2020). Actual pK_{SP} values are not comparable because of the varying stoichiometry of reaction. Affinity and the levels of bicarbonate and phosphate present in the mainline mean that a significant proportion precipitates in the mainline and passes through to biosolids treatment. During anaerobic digestion, metals reduce, and generally bind strongly with sulfides to form metal sulfide, with, generally very high pK_{SP} values, generally far higher than that of Fe (Stumm & Morgan, 1996) (i.e., outcompete Fe for sulfide). This can be leveraged for heavy metal removal from wastewater (which contaminates biosolids product). Heavy metal concentrations are generally low enough in domestic wastewater that the impact on anion availability is minimal. However, if partitioning in the mainline is important, or heavy metal removal is required (particularly from industrial streams), the chemistry is fairly straightforward, and compatible with a generalised physicochemical approach (Maharaj *et al.*, 2019).

Silica (SiO_2) is abundant in wastewaters as a particulate solid from mineral, artificial, and natural sources, and is used in wastewater treatment as an adsorbent and surface tension modifier.

The general dissolution reaction is

$$SiO_{2(s)} + 2H_2O \leftrightarrow Si(OH)_4 \quad (pK_{SP} \sim 2.7 - 3.7) \tag{3.8}$$

With the $Si(OH)_4$ being silicic acid with pK_a values of 9.46, 12.56, and 12.57 (Stumm & Morgan, 1996). The base forms of silicic acid only form at high pH, such that solubility increases as pH increases.

Forms can be divided into crystalline silica (natural silica including quartz), and amorphous silica. Amorphous silica can be further divided into reactive (silica precipitate, sodium silicate), and non-reactive (e.g., glass). Solubility of particulate forms varies, but is in the 2–3 mg/L range for crystalline silica, and 20–40 mg/L for amorphous silica at pH 7 (Braunstein, 2010). This is driven partly by the different pK_{SP} values of 3.7 and 2.7 for crystalline and amorphous silica, respectively. Crystallisation and dissolution of crystalline forms, and stable amorphous forms (including glass) is extremely slow, such that this only occurs at high temperature, pH (or depending on the form), or both.

Silica chemistry is generally not important to domestic wastewater treatment, as pH is not high enough to allow soluble silica in substantial amounts. Silica is an important potential foulant in membrane filtration, particularly reverse osmosis, where residual soluble silicic acid is concentrated above the precipitation threshold, and the silica is not removed by acid washing (which removes acid-soluble calcium and magnesium precipitates). It is also important in industrial wastewaters where pH is high, and sodium silicates are used, or where silica is released from piping, due to high temperature and pH (e.g., paper manufacturing, casting, boiler circuits).

3.3 MODELLING APPROACH

3.3.1 General framework (excluding mainline iron/aluminium–phosphorus complexes)

With the exception of iron and aluminium phosphate adsorption on hydroxides in wastewater treatment mainlines (see Section 3.4.2), the precipitation chemistry is largely consistent, given different reactants and product crystal types. A general kinetic-limited approach with parallel liquid-phase equilibrium calculations is widely applicable. Different modelling approaches have been proposed (Kazadi Mbamba *et al.*, 2015a; Lizarralde *et al.*, 2015; Musvoto *et al.*, 2000a, 2000b). In all these approaches, precipitation kinetics is based on SI, in the different ways that SI can be expressed (Table 3.1). While in Musvoto *et al.* (2000a), the rate of precipitation only depends on SI, Kazadi Mbamba *et al.* (2015a) include the effect of precipitate concentration and Lizarralde *et al.* (2015) consider the delay of nucleation, the influence of crystal seeding and TSS concentration. Other modelling approaches, more focused on the design and optimisation of precipitation technologies, include population balance models, where the mechanisms of precipitation are described in detail and the evolution of the particle-size distribution is tracked (Elduayen-Echave *et al.*, 2021; Galbraith & Schneider, 2014).

For the model description in this report, the general framework is provided in Kazadi Mbamba *et al.* (2015a). SI is used to indicate degree of supersaturation:

$$SI = \log\left(\frac{IAP}{K_{SP}}\right) \tag{3.9}$$

This is not used directly in precipitation calculations, but is a useful output from the model to assess competition between precipitants. In some cases, it may be raised to the power of $1/n$ as in equation (3.11), but is generally used as it is, as it provides a comparable value across multiple minerals for precipitation initiation. It is recommended that *SI* calculation and an initial screening of potential minerals be undertaken to reduce the size of the problem. The PCM plant-wide model by default considers calcite (X_{CaCO_3}), aragonite ($X_{CaCO_{3a}}$), amorphous calcium phosphate ($X_{Ca_3(PO_4)_2}$), hydroxyapatite ($X_{Ca_{10}(PO_4)_6(OH)_2}$), octacalcium phosphate ($X_{Ca_8H_2(PO_4)_{6.5}H_2O}$), brushite ($X_{CaHPO_4}$), struvite ($X_{MgNH_4PO_4}$), newberyite ($X_{MgHPO_4}$), magnesite ($X_{MgCO_3}$), k-struvite ($X_{KNH_4PO_4}$), vivianite ($X_{Fe_3(PO_4)_2}$) and iron sulfide ($X_{FeS}$). Vivianite ($X_{Fe_3(PO_4)_2}$) may also be considered using this approach (in the anaerobic digester), but mainline iron and aluminium phosphate must be addressed using an adsorption model (see Section 3.4.2).

The supersaturation state for the forward precipitation reaction is the supersaturation variable (σ) for a given precipitant:

$$\sigma = \left(\frac{\mathrm{IAP}}{K_{\mathrm{SP}}}\right)^{1/n} - 1 \tag{3.10}$$

where n is the number of reactants (not including water). For calcite ($CaCO_3$), this becomes

$$\sigma_{\mathrm{CaCO_3}} = \left(\frac{\{\mathrm{Ca}^{2+}\}\{\mathrm{CO_3}^{2-}\}}{K_{\mathrm{SP,CaCO_3}}}\right)^{1/2} - 1 \tag{3.11}$$

The rate is then an nth-order kinetic implementation of this, including the impact of seed (X_{precip}):

$$r_{\mathrm{precip}} = k_{\mathrm{cryst}} X_{\mathrm{precip}} \sigma^{n} \tag{3.12}$$

The seed can be placed in the influent, or the initial state set at residually low levels (e.g., 1×10^{-6} M) to ensure that forward precipitation can occur. Below SI = 1 dissolution of this will occur, but the model will continually simulate diminishing residue to allow forward precipitation. Due to numerical resolution in differential equation solvers, this may fluctuate around zero (but never be exactly zero), and precipitate will accumulate rapidly under supersaturation conditions.

Where the DAE approach is taken from Chapter 2 (the recommended implementation) the rate equation acts on the total component (e.g., for calcite this is S_{IC} and S_{ca}) with the activities $\{\mathrm{Ca}^{2+}\}$ and $\{\mathrm{CO}_3^{2-}\}$ calculated by convergence of the algebraic equations.

3.3.2 Chemical phosphorus removal

For phosphorus removal with iron or aluminium, the model complexation and adsorption model as developed in SUMO has been used (Hauduc *et al.*, 2015) as implemented by Solon *et al.* (2017) and validated by Kazadi Mbamba *et al.* (2019). This has been applied to iron precipitation only, but is equally applicable to aluminium phosphate (with modified K_{sp} parameters, but adsorption and kinetic parameters appear applicable).

Essentially, competitive metal hydroxide and phosphate precipitation are modelled by equation (3.12), with reversible adsorption of $H_2PO_4^-$ and HPO_4^{2-} on HFO/HAO based on the activities of these ions.

3.4 CASE STUDIES

This section presents the results of two-research-oriented case studies in which the precipitation model was tested. The first study is a lab-scale test of multiple precipitates from a synthetic solution (Kazadi Mbamba *et al.*, 2015a). The second example includes a full-scale model validation and model-based analysis of a domestic wastewater treatment plant (Flores-Alsina *et al.*, 2016). These examples are relatively simple case studies that primarily show the use of a precipitation model to predict the formation of relevant minerals.

3.4.1 Case study 1: Ca–Mg–PO_4 (Kazadi Mbamba *et al.*, 2015a)

During titration, a sample is taken, and acid or base added from a high or low pH, respectively (downward or upward titration). A downward titration test is generally useful for determining acid-soluble mineral dissolution, while an upward titration is useful for determining acid-soluble mineral precipitation. Titration allows evaluation of rate and stoichiometry as pH increases to the point where the limiting reactant (anion) becomes available. This example is a synthetic test of the competition of various major divalent cations in the presence of phosphate, in the absence of inorganic carbon. The laboratory titration system in this example study used an auto-titrator with a pH sensor. The titration vessel used was a 1 L stirred glass crystalliser. During a titration, the test sample to which titrant was added was stirred with a 40 mm Teflon-coated magnetic bar stirrer at 300 rpm. The volume of each titrant added, the pH and the temperature were recorded over time through a LabX Light Titration Software interface. The synthetic wastewater contained magnesium (Mg), calcium (Ca), phosphate (PO_4) and inorganic nitrogen (the relevant form being ammonium, NH_4^+). This combination can precipitate as struvite ($MgNH_4PO_4{\cdot}6H_2O$), dicalcium phosphate dihydrate (DCPD, $CaHPO_4{\cdot}2H_2O$), amorphous calcium phosphate (ACP, Ca_3PO_4) and octacalcium phosphate (OCP, $Ca_4H(PO_4)_3{\cdot}2.5H_2O$) as indicated by calculated SI values.

The initial pH of a 1 L aliquot of test wastewater was adjusted to a desired value (pH 5) using concentrated sodium hydroxide (NaOH) or hydrochloric acid (HCl) solutions, with the amount of added acid or caustic recorded. A 2 M aqueous solution of NaOH (5–10 mL total) was then quantitatively added at a predetermined fixed rate and pH was continuously measured. 100 mg L^{-1} of struvite crystal seed was then added after 3 h, at a time point in the experiment where the solution was estimated (using SI values) to have transitioned from undersaturated to supersaturated conditions with respect to struvite.

The activity/pairing model as described in Chapter 2 was used for ion speciation. None of the parameters in the activity/pairing model were contestable, with only crystallisation kinetics parameters being estimated via simulation. The titration experiments were fitted with the baseline model of equation (3.12) using a non-linear optimisation technique, giving estimated values for k_{struv}, k_{DCPD} and k_{OCP} of 3.2 ± 1, 0.76 ± 0.69 and 1.63 ± 0.49 h^{-1} for struvite, DCPD and OCP, respectively. Figure 3.5 presents the fit together with the experimental data and shows that the baseline model fit the titration experiment effectively.

An increase in pH (Figure 3.5A) led to formation of precipitates, with a corresponding decrease in soluble phosphate (Figure 3.5B) and calcium (Figure 3.5E). This is related to the formation of a calcium phosphate mineral. Through non-linear regression, the model suggested DCPD, which became supersaturated and precipitated at around 1.8 h. As the pH continued to increase with time (for around 3.5 h), further decreases in calcium, magnesium, inorganic nitrogen and phosphate were observed, suggesting that other minerals became supersaturated and precipitated, such as struvite and OCP. When two or more precipitates formed, competition between the minerals was well described inherently by the model. In this case, key competitive minerals appeared to include struvite and calcium phosphate (model indicated DCPD and OCP). The competition between these multiple minerals caused the complex interactions between pH, phosphates, and relevant cations.

3.4.2 Case study 2: application of the PCM to a full-scale municipal WWTP (Kazadi Mbamba *et al.*, 2016)

This case study examines the application of the physicochemical model to a full-scale system. The objective of the study was to assess the ability of the PCM integrated within industry-standard models to describe the process performance at a typical biological nutrient removal plant using the plant routine dataset. A step-wise methodology was used in this study and included the following steps: (1) collection of plant design and operational dataset as well as carrying out additional measurement campaign to augment data suitable for the PCM; (2) the Benchmark Simulation Model no 2 (BSM2) extended with the PCM framework was configured in accordance with the design parameters of the full-scale plant; (3) the resulting model was calibrated by adjusting the relevant key kinetic and stochiometric parameters after comparing the measured and modelled values; (4) the model was validated with a different set of data to examine how well the predicted results match the plant performance in terms of treated effluent water quality, solids production and biogas production; (5) the calibrated model was then used to run scenario analysis for nutrients recovery.

The modelled full-scale WWTP, located in South East Queensland, was designed for removal of organic matter (COD), nitrogen (N) and phosphorus (P) from domestic wastewater of approximately

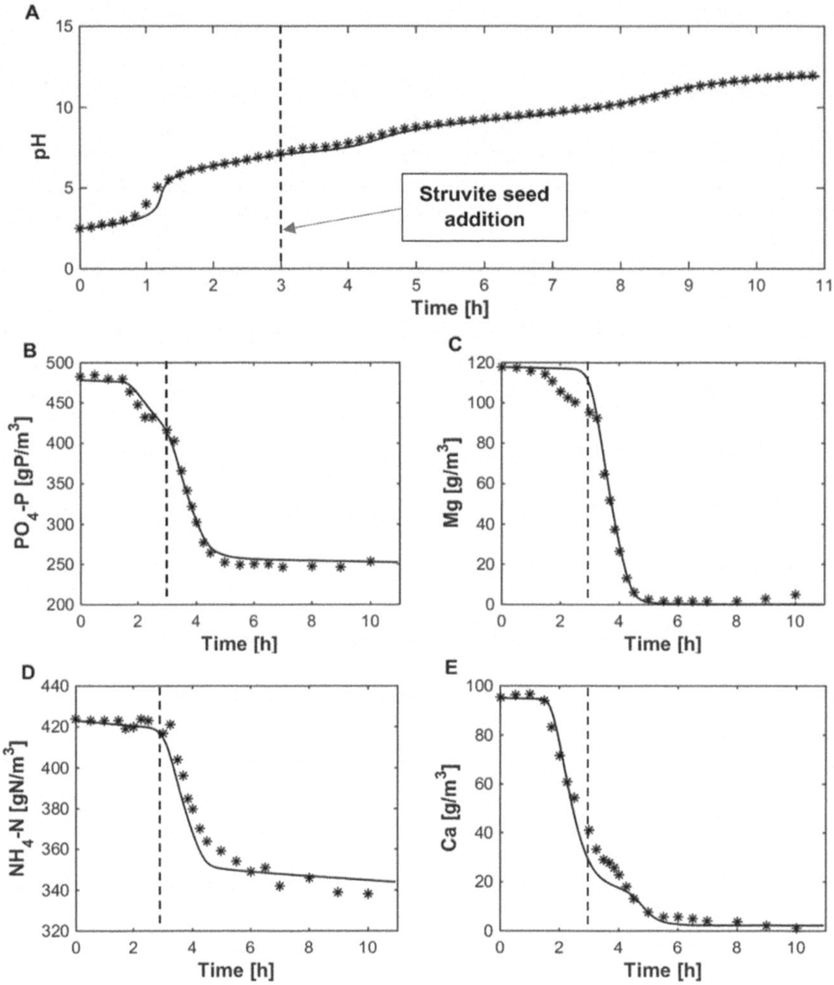

Figure 3.5 Experimental data vs. model prediction for the dynamic titration of a synthetic wastewater. Struvite crystal seeds were added after 3 h of titration when the solution became supersaturated with respect to struvite.

750 000 equivalent persons (EP). Over the period of this study the average flowrate of influent wastewater was 143 500 $m^3\ d^{-1}$ and the influent discharge loads were approximately 100 000 kg COD d^{-1}, 5600 kg N d^{-1} and 730 kg P d^{-1}. The plant-wide configuration of the full-scale wastewater treatment plant was developed based on BSM2 but expanded with an improved physicochemical model consisting of an equilibrium approach and ion pairing, kinetic minerals precipitation.

The model-based analysis of the full-scale plant consisted a check of its capability to describe the water and sludge lines at different points along the treatment process at steady state. Figure 3.6 presents the modelled and measured data for selected variables under steady-state conditions. Default parameters were used for ASM2d (Henze *et al.*, 2000), while default and new parameters, describing phosphorus model, were applied for ADM1 in the plant-wide model (Batstone *et al.*, 2002a; Flores-Alsina *et al.*, 2015). The model parameters for precipitation of minerals were adopted from a batch study using digestate from the wastewater treatment plant under study (Kazadi Mbamba *et al.*, 2015b).

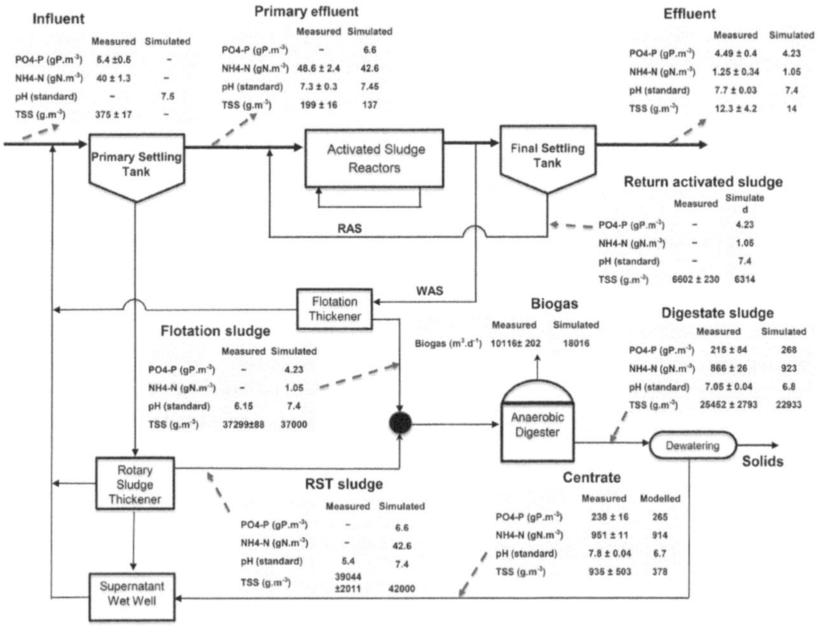

Figure 3.6 Steady-state comparison between the model prediction and measured data for representative streams and variables across the wastewater treatment plant. (*Source*: Reproduced from Kazadi Mbamba *et al.*, 2016).

The parameters defined in BSM2 were originally for clarification and settling models (Gernaey *et al.*, 2014) as implemented in Solon *et al.* (2017).

With these rate kinetic and stoichiometric parameters, there was a reasonably good agreement between the full-scale experimental dataset and simulated results for different water and sludge streams around a wastewater treatment plant under steady-state conditions. The differences between the measured and modelled data throughout the plant ranged from 4 to 15%, while the relative errors for ammonia nitrogen ranged between 4% and 38%. The model predicted a lower value (0.76 g m^{-3}) than the measured one (1.23 g m^{-3}). The relative error for the measured and predicted pH in the primary effluent, effluent and digester were 2%, 4% and 3%, respectively. However, notable discrepancies were observed after separation processes. For example, the relative errors of the measured and predicted were observed in the DAF thickened sludge (20%) and RST thickened sludge (37%). High difference between measured and measured pH could be explained by the fact that major processes that affects the pH and the concentrations of the soluble component are not included in the sedimentation, clarification, thickening and dewatering models. To obtain a plant-wide model that is realistic thickening and thinning process would probably need to be applied not to the solids but also on the soluble components in the clarifying and thickening models. In addition, a 14% error difference was observed in the centrate after dewatering. The increase in pH in the centrate (reject water) was an indication of inadvertent CO_2 stripping in the pipes and during dewatering, known as causes of inorganic scale formation. This could be corrected by adding a stripping model, and if required a precipitation model before or within the dewatering unit to simulate inadvertent gas stripping and spontaneous minerals precipitation.

The steady-state results of this case study illustrated the usefulness of a fully fledged PCM integrated within biological models that is able to capture the mechanisms of precipitation of minerals and can provide detailed insight into the interactions between major components such as iron species, sulfur and phosphorus within a biological nutrient removal/recovery at full scale (Kazadi Mbamba *et al.*, 2019). Consequently, such an integrated model could be used for control strategies to optimise the secondary effluent quality in terms of N and P, minimise inadvertent spontaneous mineral precipitation while maximising nutrient recovery (Gernaey *et al.*, 2014). For example, operational and control strategies to avert mineral precipitation from taking place in the anaerobic digesters include operating digesters at a depressed pH (with some decrease in biological activity), to ensure conditions are below the saturation points of minerals (Doyle & Parsons, 2002; van Rensburg *et al.*, 2003). It is therefore recommended that mineral precipitation be considered in the digesters of the plant-wide model. The routine monitoring data were augmented through additional wastewater sampling and measurement to provide enough information for the model-based analysis of the physico-chemistry during wastewater treatment. The use of a PCM with biological models needs additional influent characterisation of soluble chemical components such as Al, Ca, Cl, Fe(II), Fe(III), TIC, sulfide, K, Mg, Na and SO_4. However, in general, most full-scale wastewater treatment plants perform routine measurements which may also include limited measurements of soluble chemical components. However, reliable model estimates of the pH and minerals precipitation would need additional measurements to target all the relevant components that play a significant role in the physico-chemistry module.

3.5 CONCLUSIONS AND PERSPECTIVE

As noted above, precipitation has been investigated along multiple lines, by multiple groups, generally focusing on different mineral groups. The approach outlined here is consistent across conventional treatment units, is largely equilibrium driven in terms of including a supersaturation driving force based on speciation in the aqueous liquid phase, and while focused on wastewater treatment, is broadly applicable to the artificial water cycle. Precipitation is, however, a kinetic process, with widely varying kinetic coefficients, and this is the largest driver for uncertainty in simulation of precipitation reactions. As noted in the minerals review, minerals can be divided according to their formation kinetics into very fast (metal sulfides – min^{-1}), fast (e.g., struvite 1–5 < h^{-1}), moderate (e.g., calcium carbonate <1 h^{-1}), and slow (HAP–$\sim d^{-1}$). An initial screening should be done for the relevant cation–anion complexes to be included based on calculated saturation indices, and kinetic coefficients initially set to order of magnitude levels for each mineral. The results from simulations can then be compared against measured values. If the simulations (all cations and anions) deviate consistently from measured results across all cations or anions, the mineral selection is correct, but the kinetics are incorrect (for some or all precipitates) and can be readily adjusted. If a balanced group of cations and anions deviates consistently, one or more kinetic coefficient is incorrect but can also be readily adjusted. If a single cation or anion is limiting the reaction in the model, the mineral mix should be changed.

The focus of precipitation modelling to date has been on description of relevant ions (particularly phosphorus) in a plant-wide context, and this is a challenging test for precipitation models due to the complex and often competing precipitates available. However, the clearly emerging application for precipitation models, and where it is critically important is for nutrient and resource recovery. The equilibrium-kinetic model presented here is clearly effective for struvite precipitation (e.g., Kazadi Mbamba *et al.*, 2015a), and appears to be applicable for other cations (e.g., calcium, magnesium). A key challenge for nutrient recovery trains is likely to be where high-strength acid or base streams are used to dissolve and recover nutrients, where non-ideality becomes more critical (as addressed in Chapter 2). However, the use of a common model will allow nutrient recovery strategies to be effectively assessed using an integrated, plant-wide approach.

doi: 10.2166/9781780409832_0039

Chapter 4
Gas–liquid transfer

Sylvie Gillot[1], Eveline Volcke[2], Izarro Lizarralde[3], Youri Amerlink[4] and Damien J. Batstone[5]

[1]IRSTEA, UR REVERSAAL, F-69626, Villeurbanne Cedex, France
[2]Department of Green Chemistry and Technology, Ghent University, Coupure links 653, B-9000 Gent, Belgium
[3]University of Navarra, Manuel Lardizabal 15, 20018 Donostia/San Sebastián
[4]Aquafin NV, Belgium
[5]Australian Centre for Water and Environmental Biotechnology, The University of Queensland, St Lucia, Brisbane, QLD 4072, Australia

In wastewater treatment and recovery, many reactions depend on mass transfer processes between a gaseous and a liquid phase. This is a multi-film diffusive process as further discussed below.

Historically, the supply of oxygen in aerobic bioreactors has been first described and included into biokinetic models, as a key process and due to its major contribution to the energy expenditure of these systems. Increasing concerns on greenhouse gas emissions (N_2O, CH_4) and the need for a better description of acid–base equilibria more recently called for a generalised approach to represent mass transfer through an interface. Dissolved gas may be supplied from the gas phase, in which case, the main direction of transfer is gas to liquid (solubilisation or adsorption), or from the liquid phase, often due to liquid-phase biological or chemical reactions (in which case, the main direction of transfer is liquid to gas – volatilisation or stripping). In the latter case, gas production may be wholly due to liquid–gas transfer. Dissolved gas may also be a biochemical inhibitor (e.g., O_2 for algae, or H_2S).

This chapter first presents the kinetic expressions that have been developed to describe mass transfer through an interface, and the models used to characterise mass transfer coefficients. Gas–liquid transfer is kinetically limited, and this is also impacted by temperature and solvent concentration matrix effects. This chapter also discusses these impacts, towards formulation of the kinetic expression of gas transfer used in the PCM, and widely elsewhere.

4.1 FUNDAMENTALS OF MASS TRANSFER

This section first discusses fundamentals of diffusion within one phase (4.1.1) followed by the principles of mass transfer through an interface (4.1.2).

4.1.1 Diffusion within one phase: the Fick's first law

Molecular diffusion is the mechanism of transfer of a constituent within a phase through the movement of molecules of this constituent under the effect of a concentration gradient. The diffusion process is governed by Fick's first law (equation (4.1)), which expresses that the mass flux of i (J_i) diffusing in the direction z, is proportional to the concentration gradient of i in the relevant phase:

$$J_i = -D_i \frac{\partial [S_i]}{\partial z} \tag{4.1}$$

where J_i is the mass flux of i (g s^{-1} m^{-2}), D_i is the diffusion coefficient of i in the medium (m^2 s^{-1}) and $\partial [S]_i / \partial z$ is the concentration gradient of i in the direction z (g m^{-3})

Assuming a unidirectional diffusion through a stationary film of thickness δ where the concentration gradient shows a linear profile $\left((\partial[S_i]) / \partial z\right) = \left((\Delta[S_i]) / \delta\right)$, the mass flux is written as

$$J_i = \frac{D_i}{\delta} \Delta[S_i] \tag{4.2}$$

The mass transfer coefficient (k) is defined as the ratio of film diffusivity to the film thickness (δ):

$$k = \frac{D_i}{\delta} \tag{4.3}$$

and therefore J_i is written as the product of the transfer coefficient by the concentration gradient:

$$J_i = k \Delta[S_i] \tag{4.4}$$

4.1.2 Mass transfer through an interface: two-film theory

Gas–liquid or liquid–gas mass transfers are driven by concentration gradients between the liquid and the gas phases (adjusted for gas solubility – see Section 4.1.3). Considering that the two phases are separated by an interface, the two-film theory proposed by Lewis and Whitman (1924) is based on the assumption that a thin layer of stagnant fluid exists on each side of the interface. Mass transfer within those layers is governed by diffusion, while convection is neglected. Beyond the thin layers, the convection eliminates concentration gradients. Figure 4.1 represents the two-film theory in the case of the absorption of a gas into a liquid, in which the interface is surrounded with layers of stagnant gas (thickness δ_G) and stagnant liquid (thickness δ_L).

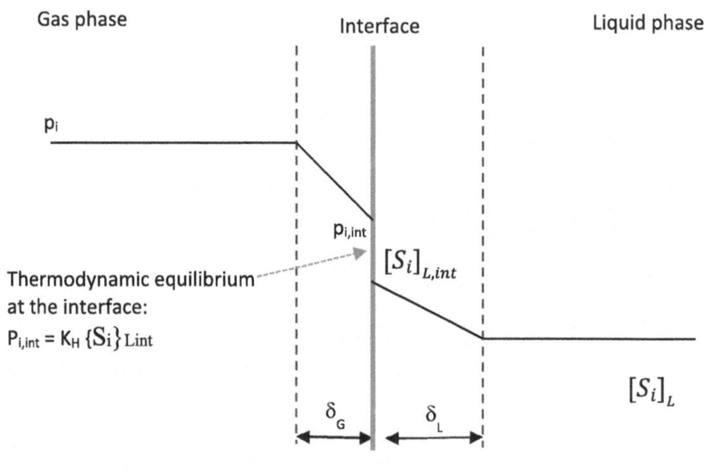

Figure 4.1 Two-film theory – adsorption of a gas in a liquid phase.

Gas concentration can be represented by gas molar concentration ($[S_i]_G$) (or molar activity for non-ideal gas mixtures), or partial pressure (p_i). Following equation (4.4), the mass flux in the gas film, J°_G (g m^{-2} s^{-1}), using gas partial pressure (p_i) instead of gas molar concentration is written as

$$J^\circ_G = k_G(p_i - p_{i,\text{int}}) \tag{4.5}$$

and the mass flux in the liquid film, J°_L (g m^{-2} s^{-1}), is expressed as

$$J^\circ_L = k_L([S_i]_{L,\text{int}} - [S_i]_L) \tag{4.6}$$

in which k_G denotes the gas-side mass transfer coefficient (g m^{-2} s^{-1} bar^{-1}), k_L is the liquid-side mass transfer coefficient (m s^{-1}), p_G is the partial pressure of the solute in the gas phase (bar), p_{Gint} is the gas-side partial pressure at the interface (bar) $[S_i]_{L,\text{int}}$ is the liquid-side concentration at the interface (g m^{-3}), and $[S_i]_L$ is the solute concentration in the homogenous liquid phase (g m^{-3}).

Note that the units of the gas-side mass transfer coefficient (k_G) have been adapted to the units usually in usage in the wastewater treatment field. Special care must therefore be paid when using literature data in the given equations (see Section 4.2.3).

At the interface, thermodynamic equilibrium is established, which can be described through Henry's law described in the following.

4.1.3 Henry's law

At equilibrium, the gas has a specific solubility limit (in the liquid) and conversely, the solute exerts a gas pressure (Stumm & Morgan 1996). This is termed Henry's law, which states that the activity of the dissolved gas i $\{S_i\}_l$ is proportional to its activity in the gas phase $\{S_i\}_g$ (or concentration – i.e., pressure for an ideal gas). There are many ways to express this proportionality, such as using the dimensionless Henry's law coefficient H_i representing the solubility of the gas:

$$\text{H}_i = \frac{\{S_i\}_L}{\{S_i\}_G} \tag{4.7}$$

Alternatively, Henry's law coefficient can also be expressed as volatility rather than solubility:

$$K_{H,i} = \frac{1}{\text{H}_i} = \frac{\{S_i\}_G}{\{S_i\}_L} \tag{4.8}$$

In case the gas is assumed to be an ideal mixture (this may not be the case at elevated pressure or temperature, but is reasonable for common mixtures at room temperature) gas concentration can be expressed as pressure, and equation (4.7) reduces to the more common form of gas–liquid equilibria:

$$\text{H}_i = \frac{\{S_i\}_L}{p_i} \tag{4.9}$$

in which p_i is the gas pressure, and H_i is expressed in mol L^{-1} bar^{-1}.

Similarly, equation (4.8) expresses the gas volatility as

$$p_i = K_{H,i}\{S_i\}_L \tag{4.10}$$

Many solution effects influence gas–liquid transfer, including change in solubility with temperature (which can be described using the van't Hoff equation), changes in activity of S_i with changes in ionic strength, acid–base and other ion pairing of component i impact its effective concentration and availability with respect to gas–liquid transfer. These are all as described in Chapter 2. Solvent activity and the presence of other ions impact the activity of the soluble gas in the liquid. Gas and liquid activities (or gas pressure and liquid activity for an ideal mixture) interact strongly and bidirectionally with precipitation and acid–base transport, acting as a source and sink for soluble gases, which may, or may not be acid-base active. As such, gas–liquid transfer is linked with precipitation and acid-base transport reactions.

It is important to note that *activities* govern the gas–liquid equilibrium as such (equation (4.7)), whereas gas–liquid mass transfer rate is induced by concentration gradients and therefore is governed by *concentrations* (i.e., liquid concentrations $[S_i]$ and gas partial pressures p_i, cf. equations (4.5) and (4.6)).

Only non-ionic species participate in gas transfer, and the effect of increased ionic strength is to increase activity (see Table 2.3). This results in salting out of solutes, and an increase in activity will result in a lower equilibrium liquid concentration (for a given gas partial pressure) or vice versa.

As discussed Section 2.2, activity corrections can be applied by explicitly using activity ($\{S_i\}$) or by using concentration ($[S_i]$) together with an activity modified Henry's volatility $K_{H,i}$'. Common practice in wastewater modelling is to use concentration uniformly and modify (effectively) the Henry's volatility (commonly by determining saturation concentration ($[S_i]_L^*$) as a function of activity coefficient γ_i:

$$[S_i]_L^* = \frac{p_i}{K_{H,i}\gamma_i} \tag{4.11}$$

This effectively describes for example the decrease in liquid saturation concentration for oxygen (partial pressure is fixed) as ionic strength increases.

Likewise, an equilibrium vapour pressure can be determined as a function of activity or liquid concentration as

$$p_i^* = K_{H,i}\{S_i\}_L = K_{H,i}\gamma_i[S_i]_L \tag{4.12}$$

This concept is used further later in the chapter, but the theory is explained by using the explicit form of species activity, consistent with the remainder of this book.

4.1.4 Two resistance model and its simplifications

In development of the two film model, an ideal mixture is assumed in the gas, while the liquid is non-ideal.

As shown in Figure 4.1, thermodynamic equilibrium is reached at the interface:

$$p_{i,\text{int}} = K_H\{S_i\}_{\text{int}} \tag{4.13}$$

This is the only part impacted by non-ideality, in the gas or the liquid, but generally liquid.

Because of the difficulty to measure the concentrations/activities at the interface (p_{Gint} and C_{Lint}), an overall mass transfer, denoted by k_{ov}, are introduced. This coefficient is based on the difference between the bulk concentration in one phase and the concentration that would be in equilibrium with the bulk concentration in the other phase ($[S_i]_L^*, p_i^*$) as defined in equations (4.11) and (4.12). The respective fluxes of transferred solute in the gas phase and the liquid phase are written as

$$J^\circ = k_{\text{ov,G}}(p_i - p_i^*) \tag{4.14}$$

$$J^\circ = k_{\text{ov,L}}([S_i]_L^* - [S_i]_L) \tag{4.15}$$

where $k_{\text{ov,G}}$ overall mass transfer coefficient expressed in gas-side units (g m^{-2} s^{-1} Pa^{-1}), and $k_{\text{ov,L}}$ is the overall mass transfer coefficient expressed in liquid side units (m s^{-1}). Note that the value of the liquid-side overall mass transfer coefficient $k_{\text{ov,L}}$ is independent of the mass units used for expressing the liquid-phase concentrations (molar vs. mass) as well as the gas-phase concentrations (pressure or concentrations).

To establish a relation between the gas-side overall mass transfer coefficient $k_{\text{ov,G}}$ and the gas-side and liquid-side transfer coefficients k_G and k_L, equation (4.14) is rearranged to incorporate the interface:

$$J^\circ = k_{\text{ov,G}}(p_i - p_i^*) \rightarrow J^\circ = k_{\text{ov,G}}((p_i - p_{i,\text{int}}) + k_{\text{ov,G}}(p_{i,\text{int}} - p_i^*)) \tag{4.16}$$

From equations (4.11) and (4.10):

$$J^\circ = k_{\text{ov,G}}(p_i - p_{i,\text{int}}) + k_{\text{ov,G}}(K_{H,i}\gamma_{i,\text{int}}[S_i]_L - K_{H,i}\gamma_i[S_i]_l) \tag{4.17}$$

Assuming $\gamma_{i,\text{int}} \sim \gamma_i$ and substituting equations (4.14) and (4.15) into equation (4.17) results in the following:

$$J^\circ = k_{\text{ov,G}}\frac{J^\circ}{k_G} + k_{\text{ov,G}}K_H\gamma_i\frac{J^\circ}{k_L} \tag{4.18}$$

Manipulation of this equation results in the following:

$$\frac{1}{k_{\text{ov,G}}} = \frac{1}{k_G} + \frac{K_H\gamma_i}{k_L} \tag{4.19}$$

The liquid-side overall mass transfer coefficient $k_{\text{ov,L}}$ can be related to the gas-side and liquid-side transfer coefficients k_G and k_L, by rearranging, Equation (4.15) in a similar way, resulting in (4.20)

$$\frac{1}{k_{\text{ov,L}}} = \frac{1}{k_L} + \frac{1}{\gamma_i K_H k_G} \tag{4.20}$$

where K_H is the Henry's law constant (bar L mole^{-1}), $1/\gamma_i K_H k_G$ resistance to mass transfer in the gas film (s m^{-1}), and $1/k_L$ is the resistance in the liquid film (s m^{-1})

4.1.4.1 Effect of gas solubility

From equations (4.19) and (4.2), it is clear that mass transfer resistance is located in both the liquid side and in the gas side. Considering constituents' solubility, three different cases may be considered:

(i) Very soluble gases (low K_H): in this case (e.g. NH_3), the film created in the water phase offers low resistance to diffusion compared to the gaseous phase and consequently it can be neglected. In this case, equation (4.19) is reduced to

$$k_{\text{ov,G}} = k_G \tag{4.21}$$

Therefore, the absorption rate is controlled by diffusion in the gas phase. By substituting equation (4.21) into equation (4.14), the mass flux can be written as follows:

$$J^\circ = k_G(p_i - p_i^*) \tag{4.22}$$

Substituting various expressions for Henry's law (4.11, 4.12) into equation (4.22) and equating into equation (4.15) results in the following:

$$J^\circ = k_G(K_{H,i}\gamma_i[S_i]_L^* - K_{H,i}\gamma_i[S_i]_L) = k_{\text{ov,L}}([S_i]_L^* - [S_i]_L) \tag{4.23}$$

Resulting in equation (4.24)

$$k_{\text{ov,L}} \cong k_G\gamma_i K_H \tag{4.24}$$

(ii) Poorly soluble gases (high K_H): with gases of low solubility (e.g. O_2, CO_2, CH_4), the solute diffuses slowly through the liquid film. In this case, equation (4.20) is reduced to

$$k_{ov,L} \cong k_L \tag{4.25}$$

which expresses that the resistance to transfer is located in the liquid phase. Hence, the expression for the mass flux (equation (4.15)) becomes

$$J^\circ = k_L([S_i]_L^* - [S_i]_L) \tag{4.26}$$

which can also be written as

$$J^\circ = k_L\left(p_i/\gamma_i K_H - [S_i]_L\right) \tag{4.27}$$

(iii) Gases with intermediate solubility: both the gas-phase and liquid-phase mass transfer resistance have to be considered. The overall mass transfer coefficient $k_{L,ov}$ is a function of both k_L and k_G and is obtained by rearranging equation (4.20) as

$$k_{ov,L} = \frac{k_L k_G}{k_G + (k_L/\gamma_i K_H)} \tag{4.28}$$

Hence, the expression for the mass flux (equation (65)) becomes

$$J^\circ = \frac{k_L k_G}{k_G + \frac{k_L}{\gamma_i K_H}}\left(\frac{p_i}{\gamma_i K_H} - [S_i]_L\right) \tag{4.29}$$

4.1.5 Mass transfer rate

Considering an interface between the gas and the liquid phases with an area A (m²), the total flux $r_{t,i}$ (g m^{-3} s^{-1}) through the interface (gas to liquid) can be written as follows:

$$\begin{aligned} r_{t,i} &= \frac{J_i A}{V_L} = k_{ov,L,i}\frac{A}{V_L}\left([S_i]_L^* - [S_i]_L\right) \\ &= k_{ov,L,i}a\left([S_i]_L^* - [S_i]_L\right) = k_{ov,L,i}a\left(\frac{p_i}{\gamma_i K_{Hi}} - [S_i]_L\right) \end{aligned} \tag{4.30}$$

$a = A/V_L$ (m² m^{-3}) is the interfacial area and V_L (m³) the volume of the liquid phase. This can be rewritten in terms of activity as

$$r_{t,i} = k'_{ov,L,i}\, a\left(\frac{p_i}{K_{Hi}} - \{S_i\}_L\right) \tag{4.31}$$

where $k'_{ov,L,i}$ is adjusted for activity (this adjustment is minor – the critical modification is to the equilibria, as the increased vapour pressure (or decreased solubility) of the gas/liquid system in response to an increase in activity needs to be accounted for. The preferred form of rate equation is further discussed in the conclusions, but for gas → liquid, non-pairing species, liquid species concentration is recommended, while for liquid → gas, pairing species, liquid species activity is recommended.

The overall mass transfer coefficient ($k_{ov,L,i}$) is obtained from equation (4.29) for gases with intermediate solubility, and can be simplified for highly soluble (equation (4.24)) or poorly soluble (equation (4.25)) gases.

Alternatively, equation (4.30) could be expressed in terms of the gas-side overall mass transfer coefficient, taking into account equation (4.19). However, the $k_{ov,L,i}$ is generally used in wastewater and other process modelling. Moreover, it has the advantage of being independent of the mass units used.

From equation (4.30), it is clear that modelling mass transfer between a gas and a liquid phase requires the determination of three components:

- The overall mass transfer coefficient $k_{ov,L,i}$
- The interfacial area a
- The concentration gradient $((p_i / (\gamma_i K_{Hi})) / -[S_i]_L)$; note that for this concentration gradient is positive in case of gas solubilisation (absorption), leading to a positive transfer rate. It is negative in case of gas volatilisation (stripping).

4.2 MASS TRANSFER COEFFICIENT VALUES

This section identifies empirical correlations that are used to estimate mass transfer coefficients. Application to different gases relevant to wastewater treatment/recovery is then proposed in the case of transfer from/to bubbles.

4.2.1 Correlations to determine mass transfer coefficient values

The mass transfer coefficients for the gas and liquid phases are defined as the ratio between the diffusion coefficient (D) and the thickness of the film (δ) (equation (4.2)). As the thickness of the film is not known, empirical correlations are typically used to determine the mass transfer coefficients (Wang *et al.*, 2005). These correlations account for the reactor geometry, the turbulence at the interface, and the properties of the two phases. Different mixing conditions and gas solubility may lead to different mass transfer coefficient values. In addition, correlations have been generally developed in ideal conditions (i.e. clean water, no solids, etc). There are specific corrections to account particularly for varying liquid matrix effects (including fundamental corrections to apparent diffusivity), and the basic correction for activity to mass transfer coefficient is discussed above. Tables 4.1–4.3 gather different equations considering either transfer from/to gas bubbles, from/to passive liquid surfaces, and in packed columns.

As far as bubbles are considered (in the presence of an active gas flow such as aeration or intense bubbling, see Table 4.1), the main distinction lies in the bubble size. For bubble diameters larger than 2.5 mm, the interface is considered as mobile and therefore variable. When smaller than 2.0 mm, the interface is hardly deformed and the bubble approximates to a sphere.

Passive liquid surfaces (Table 4.2) refer to the absence of active gas flow, such as defined by Prata *et al.* (2018) in their review.

For packed columns (Table 4.3), a review of mass transfer correlations has been made by Wang *et al.* (2005). The accuracy of these correlations usually lies around 30%, but larger errors are not uncommon (Cussler, 2009).

Most of the correlations refer to the liquid side coefficient $k_{L,i}$. only. This can be attributed to the fact that most gases of interest have a low solubility, which makes that the mass transfer resistance is located in the liquid phase. In particular, many experiments were performed with oxygen, which is a poorly soluble gas.

Note that for this section, activity correction is not included, since it generally changes mass transfer coefficients incrementally. In general, where K_H appears, activity correction should be included as $\gamma_i K_{Hi}$, and where H appears, activity correction is included as H_i/γ_i.

Table 4.1 Correlations to determine local mass transfer coefficients ($k_{L,i}$ and $k_{G,i}$) in the presence of an active gas flow creating bubbles.

Interface Created Through	Local Mass Transfer Coefficient		Reference
Bubbles with rigid interface $db < 2.0 \times 10^{-3}$ m	$k_{L,i} = 0.31\left(\frac{g\mu_L}{\rho_L}\right)^{1/3} Sc^{-2/3}$	ρ_L liquid density (kg m^{-3}) μ_L liquid viscosity (Pa s) Sc Schmidt number ($Sc = \mu_L/(\rho_L D_{L,i})$) $D_{L,i}$ solute diffusivity in the liquid (m^2 s^{-1})	Calderbank and Moo-Young (1961)
	$k_{L,i} = \frac{D_{L,i}}{d_b}(2 + 0.6Re^{1/2}Sc^{1/3})$	Re bubble Reynolds number ($Re = U_b d_b \rho_L / \mu_L$) Sc Schmidt number ($Sc = \mu_L/(\rho_L D)$) U_b bubble rising velocity (m s^{-1})	Frossling (1938)
Bubbles with mobile interface $db > 2.5 \times 10^{-3}$ m	$k_{L,i} = 2 \cdot \sqrt{\frac{D_{L,i} \cdot v_r}{\pi \cdot d_b}}$	$D_{L,i}$ solute diffusivity in the liquid (m^2 s^{-1}) d_b bubble diameter (m) v_r bubble slip velocity (m s^{-1})	Higbie (1935)
	$k_{L,i} = 0.42\left(\frac{(\rho_L - \rho_G)\mu_L g}{\rho_L^2}\right)^{1/3} Sc^{-1/2}$	ρ_L liquid density (kg m^{-3}) ρ_G gas density (kg m^{-3}) μ_L liquid viscosity (Pa s) Sc Schmidt number ($Sc = \mu_L/(\rho_L D_{L,i})$) $D_{L,i}$ solute diffusivity in the liquid (m^2 s^{-1})	Calderbank and Moo-Young (1961)
	$k_{L,i} = \sqrt{D_{L,i} s}$	s surface renewal rate (s^{-1})	Danckwerts (1951)
Bubbles	$Sh_G = \frac{k_{G,i} d_b}{D_{G,i}} = 6.58$	d_b bubble diameter (m) $D_{G,i}$ solute diffusivity in the gas (m^2 s^{-1})	Sharma and Mashelkar (1968)

4.2.2 Examples of applications

In recent papers (Lizarralde *et al.*, 2018; Vaneeckhaute *et al.*, 2018), the Higbie correlation (Higbie, 1935) is used to describe gas transfer in processes implemented for water resource recovery modelling. However, this correlation was initially developed to determine the liquid side coefficient, assuming a low solubility. Special attention must be paid to ensure the validity of this correlation (see relevant case in Chapter 6).

Table 4.2 Correlations to determine local mass transfer coefficients ($k_{L,i}$ and $k_{G,i}$) for passive liquid surface.

Interface Created Through	Local Mass Transfer Coefficient		Reference
Open channels	$k_{L,i} = \sqrt{\frac{D_{L,i}\overline{U_L}}{H}}$	$\overline{U_L}$ mean liquid velocity (m s^{-1})	O'Connor and Dobbins (1956)
Agitated vessels	$k_{L,i} \sim \left(\frac{P}{V}\right)^n$	P/V specific power input (W/m^3)	Munz and Roberts (1984)
Any surface	$k_{L,i} = 2 \cdot \sqrt{\frac{D_{L,i}}{\pi t_c}}$	$D_{L,i}$ solute diffusivity in the liquid (m^2 s^{-1}) t_c contact time (s)	Higbie (1935)

Table 4.3 Correlations to determine local mass transfer coefficients ($k_{L,i}$ and $k_{G,i}$) in packed-bed reactors.

Reference	Local Mass Transfer Coefficient	
Billet and Schultes (1993)	$k_G = cp_G \cdot \frac{a^{0.5} \cdot D_G}{\sqrt{d_h(\varepsilon - h_L)}} \cdot \left(\frac{\rho_G \cdot u_G}{a \cdot \mu_G}\right)^{3/4} \cdot \left(\frac{\mu_G}{\rho_G \cdot D_G}\right)^{1/3}$ $k_L = cp_L \cdot \left(\frac{\rho_L \cdot g}{u_L}\right)^{1/6} \cdot \left(\frac{D_L}{d_h}\right)^{0.5} \left(\frac{u_L}{a}\right)^{1/3}$	a packing material specific surface area ($m^2\ m^{-3}$) ε void fraction of the packing (dimensionless) ρ volumetric concentration, volumetric mass (kg m^{-3}) μ dynamic viscosity (Pa.s) cp_G packing specific coefficient for the gas phase (dimensionless) cp_L packing specific coefficient for the liquid phase (dimensionless) d_h hydraulic diameter of the packing (m) d_p particle diameter (m) D_G gas diffusion constant ($m^2\ s^{-1}$) D_L liquid diffusion constant ($m^2\ s^{-1}$) g acceleration of gravity (m s^{-2}) h_L liquid hold-up of the packing (dimensionless) u_G gas velocity (m s^{-1}) u_L liquid velocity (m s^{-1})
Onda *et al.* (1968)	$k_G = 5.23aD_G\left(\frac{u_G\rho_G}{a\mu_G}\right)^{0.7}\left(\frac{\mu_G}{\rho_G D_G}\right)^{1/3}(ad_p)^{-2.0}$ $k_L = 0.0051\left(\frac{\rho_L}{\mu_L g}\right)^{-1/3}\left(\frac{u_L\rho_L}{a\mu_L}\right)^{2/3}\left(\frac{\mu_L}{\rho_L D_L}\right)^{-0.5}(ad_p)^{0.4}$	
Shulman *et al.* (1955)	$k_G = 1.95u_G\left(\frac{d_p u_G \rho_G}{\mu_G(1-\varepsilon)}\right)^{-0.36} Sc^{-2/3}$ $k_L = 25.1\left(\frac{D_L}{d_p}\right)\left(\frac{d_p u_L \rho_L}{\mu_L}\right)^{-0.36} Sc^{0.5}$	

Source: From Van der Heyden *et al.* (2016).

Van der Heyden *et al.* (2016) compared the correlations of Onda *et al.* (1968), Shulman *et al.* (1955) and Billet and Schultes (1993) to describe the mass transfer of NH_3 in a chemical air scrubber, for different types of packing material, in terms of the specific surface area and the void fraction. The correlation of Billet and Schultes (1993) was found to be the most flexible correlation and valid for different packing materials. This one was therefore incorporated in their model.

In their study of aerobic oxidation of H_2S in a trickling filter, López *et al.* (2016) found that the contribution of the gas-side mass transfer resistance was negligible for both O_2 and H_2S, and that only the liquid-side mass transfer resistance was significant. In a later study for a similar setup (López *et al.*, 2021), it was concluded that the Billet and Schultes (1993) correlation for the prediction of $k_{L,H2S}$ fitted the experimental data better than that of Onda *et al.* (1968).

4.2.3 Typical parameter values

Typical gaseous components considered in wastewater treatment/recovery applications are: O_2, O_3, CO_2, NH_3, N_2, H_2, CH_4, N_2O, NO, and H_2S. Table 4.4 summarises their characteristics in terms of Henry's law coefficients (high K_H denotes low solubility) and diffusivity into air and water. Based on diffusion coefficients and Henry's law coefficients from literature (Massman 1998; Sander 2015), values of gas-side and liquid-side gas local transfer coefficients ($k_{G,i}$ and $k_{L,i}$, respectively) were estimated considering a bubble of 3 mm diameter. Correlations from Table 4.1 were used to that purpose, that is the Higbie correlation for $k_{L,i}$ and Sharma and Mashelkar equation for $k_{G,i}$. The overall mass transfer coefficient ($k_{ov,L,i}$) was then deduced from equation (4.28) and the fraction of the total mass transfer resistance located in the liquid side ($k_{L,i}/\ k_{ov,L,i}$) was calculated $k_{L,i}/k_{ov,L,i}$ values reported in Table 4.4 that confirm the impact of gas solubility on the mass transfer discussed above. In case of low soluble gases (high $K_{H,i}$ values), the overall mass transfer is limited by the resistance in the liquid film ($k_{L,i} \approx k_{L,ov,i}$). Conversely in case of highly soluble gases, the overall mass transfer is gas-phase dominated

Table 4.4 Gas characteristics at 20°C and 1 bar and associated mass transfer coefficients calculated for a mobile bubble of 3 mm.

Gas	Diffusivity in Air[1]	Diffusivity in Water[1]	Henry's Law Coefficient[2]	Dimensionless Henry's Law Coefficient[2]	Gas-side Mass Transfer Coefficient[3]	Liquid-side Mass Transfer Coefficient[4]	Overall Mass Transfer Coefficient[5]	Fraction in the Liquid Side
	$D_{G,i}$	$D_{L,i}$	$K_{H,i}$	$K_{H,i}$	$k_{G,i}$	$k_{L,i}$	$k_{ov,L,i}$	$k_{L,i}/k_{ov,L,i}$
	$m^2\ s^{-1} \times 10^5$	$m^2\ s^{-1} \times 10^9$	$Pa\ m^3\ mol^{-1} \times 10^4$	Dimensionless	$m\ s^{-1} \times 10^2$	$m\ s^{-1} \times 10^4$	$m\ s^{-1} \times 10^4$	Dimensionless
O_2	2.07	2.10	7.06	28.97	4.54	4.62	4.62	1.00
O_3	1.64	1.76	0.73	2.98	3.60	4.23	4.21	0.99
NO	2.05	0.09	4.80	19.71	4.49	0.94	0.94	1.00
N_2O	1.63	1.77	0.36	1.47	3.58	4.25	4.20	0.99
N_2	2.03	1.65	14.51	59.51	4.46	4.10	4.10	1.00
H_2	7.00	4.21	12.46	51.12	15.35	6.55	6.55	1.00
H_2S	1.62	1.63	8.87×10^{-2}	0.36	3.55	4.07	3.91	0.96
CH_4	2.22	1.51	6.52	26.74	4.87	3.92	3.92	1.00
CO_2	1.57	1.73	0.26	1.08	3.44	4.20	4.14	0.99
NH_3	2.25	2.07	1.33×10^{-4}	5.47×10^{-04}	4.93	4.59	0.19	0.04

[1]From Massman (1998).
[2]From Sander (2015), except H_2 and H_2S from Roustan (2003).
[3]Calculated from Sharma and Mashelkar (1968), see Table 4.1.
[4]Calculated from Higbie (1935), see Table 4.1.
[5]Calculated from equation (4.28).

($k_{ov,L,i} \approx K_{H,i}\,k_{G,i}$). For intermediate $K_{H,i}$ values – around 10^3 Pa m^3 mol^{-1} both local transfer coefficients need to be considered to calculate $k_{ov,L,i}$(equation (4.28)) (Prata *et al.*, 2018).

4.2.4 Relationships between *k*-values for different gases

Once the mass transfer coefficient of a component has been determined, for a given reactor configuration and for defined operating conditions (in terms of gas and liquid velocities, temperature, pressure, etc), it can be used to calculate the mass transfer coefficients for other components. Indeed, the empirical equations from Tables 4.1 to 4.3 show that there exists a simple correlation between liquid side mass transfer coefficients of two gaseous components:

$$\Psi = \frac{k_{L,i}}{k_{L,j}} = \left(\frac{D_{L,i}}{D_{L,j}}\right)^n \tag{4.32}$$

In case the reference component (j) is a poorly soluble gas ($k_{ov,L,j} = k_{L,j}$), the ratio between overall mass transfer coefficients can be deduced from equations (4.28) and (4.32):

$$\begin{aligned}\Psi_{ov} &= \frac{k_{ov,L,i}}{k_{ov,L,j}} = \frac{k_{ov,L,i}}{k_{L,j}} = \frac{k_{L,i}}{k_{L,j}}\frac{(k_{G,i}/k_{L,i})}{(k_{G,i}/k_{L,i}) + (1/K_{H,i})} \\ &= \left(\frac{D_{L,i}}{D_{L,j}}\right)^n \frac{(k_{G,i}\,/\,k_{L,i})}{(k_{G,i}/k_{L,i}) + (1/K_{H,i})}\end{aligned} \tag{4.33}$$

The empirical coefficient n ranges from 0.1 to 1.0 with most likely values of 1/2, 2/3 and 1 according to Munz and Roberts (1984). These authors used equation (4.28) to correlate mass transfer coefficients obtained for different volatile organic compounds to the one of oxygen. They obtained an n value of 0.5, in accordance with Higbie's equation.

In case of low soluble gases (high $K_{H,i}$ values), equation (4.33) simplifies to equation (4.32) ($\Psi_{ov} = \Psi$).

Although recent references use the value of 0.5 (Lizarralde *et al.*, 2015; Vaneeckhaute *et al.*, 2018), further validations are required for the list of gases and the unit processes usually encountered in water resource recovery systems.

From a practical point of view, this result is important as not all mass transfer rates are easily observable experimentally. Equation (4.33) (equation (4.32) for highly soluble gases) enables to deduce the mass transfer coefficient for any gas from mass transfer measurements for a single gas, usually oxygen in aerobic conditions and methane, carbon dioxide, and hydrogen in anaerobic conditions.

4.3 DETERMINATION OF THE INTERFACIAL GAS–LIQUID TRANSFER AREA

The interfacial area for gas–liquid transfer a, which is required to determine the exchange rate (equation (78)), depends on geometric properties of the reactor and on physical–chemical properties of the system in case of gas injection systems. For the calculation of a, a distinction is made between passive and active gas exchange. Passive gas exchange denotes gas exchange through a surface and/or due to bubble generation through biological reactions. Active gas exchange is obtained by an active supply of gas into the reactor, through air injection or liquid projection into the atmospheric air.

In case of passive gas exchange through the reactor surface area, the interfacial area is obtained as the reactor surface area A (m^2) divided by the volume of the liquid phase, V_L (m^3).

$$a = \frac{A}{V_L} \tag{4.34}$$

In case of active gas transfer due to gas injection (e.g., subsurface aeration systems), the interfacial area is the total surface of the bubbles in contact with the liquid phase. This interfacial area is estimated using equation (4.20).

$$a = \frac{6\varepsilon_G}{(1-\varepsilon_G)d_{bs}} \tag{4.35}$$

where ε_G is the gas volume fraction (dimensionless), and d_{bs} the bubble Sauter diameter (m).

The gas volume fraction is defined by equation (4.36):

$$\varepsilon_G = \frac{V_G}{V_T} = \frac{V_G}{V_G + V_L} \tag{4.36}$$

where V_G is the volume occupied by the gas phase (m^3), V_T is the aerated volume (m^3), and V_L is the volume occupied by the liquid phase (m^3).

The bubble Sauter diameter (d_{bs}) is the weighed bubble's diameter considering the population bubble size distribution, defined as

$$d_{bs} = \frac{\sum_i ni \cdot dbi^3}{\sum_i ni \cdot dbi^2} \tag{4.37}$$

where dbi is the bubble diameter (m), ni is the number of bubbles with diameter dbi.

In specific reactors (distillation column, surface aeration systems, air scrubbers, quiescent reactors where bubbles form with gas generation, etc) the developed interfacial area is difficult to estimate and experiments under actual environmental conditions are advised to determine mass transfer coefficients. The values of interfacial area are then deduced from the measurement of overall mass transfer coefficients (k_La, k_Ga), estimating the local mass transfer coefficients (k_L, k_G) from available correlations (Tables 4.1–4.3). This is exemplified for oxygen in Section 4.5.4.

4.4 LIQUID SATURATION CONCENTRATION

To calculate the mass transfer of a component i, the saturation liquid concentration$[S_i]_L^*$ needs to be known (equation (4.30)). The value of $[S_i]_L^*$ can be determined experimentally, or using Henry's law (with activity correction) (equation (4.11)). However, saturation must consider pressure gradients, liquid-phase reactions, and non-ideality as discussed further below.

4.4.1 Gas-phase concentration gradients

In contrary to the case of oxygen in aerobic reactors, the partial pressure of the gas may not be considered as uniformly distributed within the reactor. Therefore, in active systems, enrichment and/or depletion of the gas in the bubbles may have to be considered to calculate the saturation concentration (Baeten *et al.*, 2020; Fiat *et al.*, 2019). In passive systems, depletion and/or enrichment in the off-gas has to be considered as well if the system is not open to the atmosphere. Dynamic simulation on the gas phase is therefore required to resolve this (see Recommended approach, Section 4.5).

4.4.2 Reactor height

The reactor height should also be considered to determine effective gas pressure. The results obtained with oxygen, using a saturation concentration obtained at mid-depth (Daelman *et al.*, 2014) or at one-third of the depth (Gillot *et al.*, 2005) cannot be extrapolated to other gases and other conditions. The gas-phase balance will therefore include gas pressure variations along the reactor depth. It is noted that this can be addressed through adjustment of the effective k_La value, but makes fundamental determination from bubble size problematic. It is noted that this is most important for gas-to-liquid transfer (importantly, oxygen transfer to liquid), and less important for liquid-to-gas transfer, as gas to liquid is promoted at higher hydraulic pressure (and vice versa).

4.4.3 Chemical equilibrium composition

The driving force for volatility of a dissolved gas is the concentration of the corresponding solute in the liquid phase. This implies that in case the volatile species is a weak acid or base (CO_2, NH_3, H_2S), only the concentration of the free form is relevant. For instance, for a given total ammoniacal nitrogen concentration ($S_{TAN} = [NH_3] + [NH_4^+]$), the free form is ammonia (the base), of which the relative concentration increases with increasing pH. As a result, following Henry's law, the gas-phase partial pressure of NH_3 also increases with increasing pH, implying a net decrease in solubility, that is the ratio of total ammoniacal nitrogen in the liquid phase versus gaseous NH_3. The opposite is true for the total inorganic carbon concentration ($S_{IC} = [CO_2] + [HCO_3^-] + [CO_3^{2-}]$), of which the solubility increases with increasing pH since the partial pressure of CO_2 decreases with increased pH.

In addition to the chemical equilibrium as such, solubility will also be influenced by ionic strength. An increase in ionic strength will cause an effective increase in volatility (decrease in solubility), commonly termed salting out (Stumm & Morgan 1996). This has been considered in development above. For oxygen transfer, this is accounted for by the β factor (see Section 4.5.4), which is effectively the inverse of activity ($\beta = 1/\gamma$), but noting that activity correction is equally applicable to other gases.

4.5 RECOMMENDED APPROACH

To use equation (4.30) to determine the total flux of a component through an interface, one should consider different scenarios according to reactor geometry. Two important features are considered in the following: (i) whether gas exchange is active or passive, and (ii) whether the reactor is closed or open to the atmosphere.

4.5.1 Mass transfer coefficients as a function of gas solubility: active versus passive gas exchange

As proposed by Vaneeckhaute *et al.* (2018), four scenarios may be envisaged, depending on gas solubility (high or low) and if the gas exchange is active or passive. Those scenarios are represented in Figure 4.2, as well as the main equations to be used in each case to determine the mass transfer rate according to equation (4.30). The equations proposed by Vaneeckhaute *et al.* (2018) have been modified according to literature correlations previously presented in Tables 4.1–4.3.

Note that the correlations have been developed for clean water conditions, and correction factors must be applied to represent process water conditions, unless experimental measurements are used to determined mass transfer coefficients.

In practice, it is often difficult to obtain reliable values for interfacial areas. Experimental values determined using oxygen in aerobic conditions and hydrogen in anaerobic conditions are used to deduce $k_{ov,L,i}a$ values for any gas i using equation (4.33), such as presented in Figure 4.2 for gases with low solubility.

Alternatively, $k_{ov,L,i}a$ values could be deduced from correlations obtained in similar reactors; those correlations usually link overall mass transfer coefficients to the input gas flow rate (Gillot *et al.*, 2005; Van der Lans 2000).

Baeten *et al.* (2020) propose analytical correlations to identify under which conditions and for which components the mole fraction and pressure gradients significantly affect the total transfer rate. A procedure quickly determines whether and how the vertical mole fraction and/or pressure gradients should be considered under specific conditions, was described and provided as a spreadsheet. This procedure was illustrated for typical concentrations of O_2, CO_2, CH_4, N_2O and N_2 in the liquid phase and inlet gas of an aerobic biological wastewater treatment reactor.

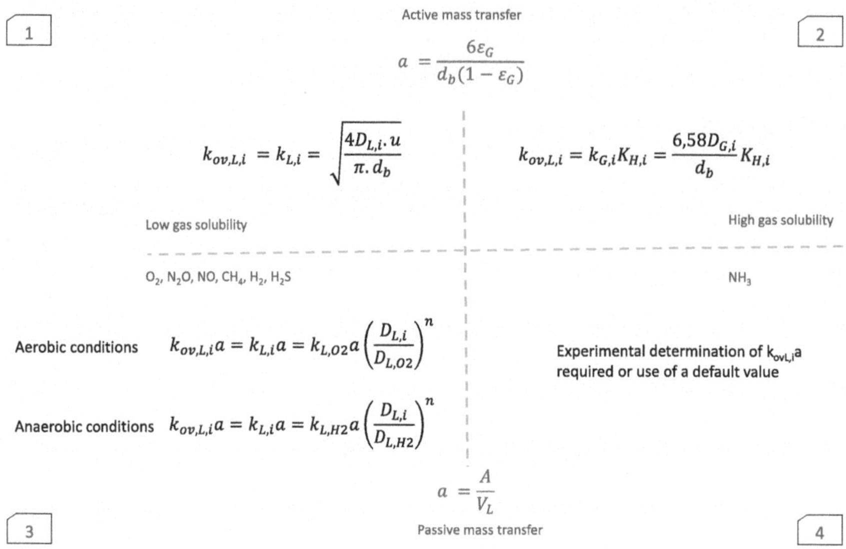

Figure 4.2 Four scenarios to calculate gas–liquid transfer coefficients.

4.5.2 Closed versus open reactors

When the reactor headspace is closed, partial pressures may vary depending on production of the various gases, and a total and component mass balance must be done in the gas phase (lumping the gas and dispersed components of the gas phase) (note that for liquid to gas, $r_{i,t}$ will be negative according to equation (4.28)):

$$\frac{dV_G p_{G,i}}{dt} = q_{G,in}p_{G,i,in} - q_{G,out}p_{G,i} + r_{i,t}V_L RT \tag{4.38}$$

where $q_{G,in}$ is the volumetric flow rate into the headspace, $q_{G,out}$ is the volumetric net flow rate out of the headspace, $p_{G,i,in}$ is the partial pressure in sweep gas inflow to the headspace, V_L is the liquid volume, V_G is the gas volume, R is the universal gas law constant (units consistent for the units of p_i and r_i), and T is the temperature in Kelvin. If headspace volume is fixed, this simplifies to

$$\frac{dp_{G,i}}{dt} = \frac{q_{G,in}}{V_G}p_{G,i,in} - \frac{q_{G,out}}{V_G}p_{G,i} + r_{i,t}\frac{V_L}{V_G}RT \tag{4.39}$$

If volume is fixed, and there is no inflow (i.e., an anaerobic digester), this simplifies to

$$\frac{dp_{G,i}}{dt} = -\frac{q_G}{V_G}p_{G,i} + r_{i,t}\frac{V_L}{V_g}RT \tag{4.40}$$

The gasflow can be determined by net gas production rate:

$$q_G = \sum \frac{-r_{i,T}V_L RT}{P_{headspace}} \tag{4.41}$$

Or determined by regulating the headspace pressure by flow through an orifice (Batstone *et al.*, 2002a):

$$q_{G,out} = k_v \sqrt{\sum p_i - P_{headspace}} \quad (4.42)$$

where $P_{headspace}$ is the headspace or downstream pressure (generally set to 1 atm). k_v is the valve coefficient, and is normally set arbitrarily high that $\Sigma p_i \sim P_{headspace}$. It is also reasonable (though not strictly correct) to make equation (4.42) linear in ΔP.

Water vapour needs to be also accounted for, normally by adding the partial pressure of water at the headspace pressure, for q_G correction in equation (4.41), or added to the Σp_i term in equation (4.42).

4.5.3 Temperature dependency

Temperature impacts both saturation concentrations and mass transfer coefficients, with gas solubility almost always decreasing with temperatures (i.e., K_H increases). The correction of Henry's law coefficient (expressed as volatility) to 20°C is given by the van't Hoff equation (e.g. Sander, 2015) (i.e., the explicit form of equation (2.5)):

$$K_{H,i}(T) = K_{H,i}(20°C) e^{\frac{\Delta H_{sol}}{R}\left(\frac{1}{T} - \frac{1}{293.15}\right)} \quad (4.43)$$

where $\Delta_{sol}H$ is the standard reaction enthalpy (J mol^{-1}), $K_{H,i}$ (20°C) is the Henry's law coefficient at the reference temperature (293.15), R is the gas constant (8.314 J mol^{-1} K^{-1}).

A theta function is commonly used in place of equation (4.43) with a reference temperature of 20°C using equation (4.44) (ASCE, 2007; Henze *et al.*, 2000):

$$k_{ov,L,i}(T) = k_{ov,L,i}(20°C)\theta^{(T-20)} \quad (4.44)$$

Values of θ reported in literature varies from 1.008 to 1.047 (ASCE 2007), 1.024 being generally adopted.

4.5.4 Application to oxygen transfer in bioreactors

Knowing the gas–liquid transfer of one gas enables the calculation of the gas–liquid transfer of other gases as described in equation (4.33). As one of the driving factors of the biological processes involved in activated sludge and due to the energy-intensive provision, oxygen transfer has been extensively studied. Standardised procedures have been developed for measuring the overall oxygen transfer coefficient ($k_{ov,L,O2}$a) and the oxygen transfer rate (OTR), amongst which the ASCE procedure (ASCE, 2007) and the European standard NFEN12255-15 (NFEN, 2004), are well accepted by practitioners. These procedures have been developed for clean water conditions. In order to calculate the actual OTR under process conditions, the standard OTR needs to be corrected for the influence of wastewater, mixed liquor, operational conditions, temperature and pressure. Due to the complex nature of these influencing factors empirical relations prevail in describing their effect on the oxygen transfer.

The total OTR is expressed as follows (adapted from Stenstrom & Gilbert, 1981):

$$r_{t,O2} = \alpha F k_{ov,L,O2} a(20°C)(\beta \tau \Omega [O_2]^*_{L,20} - [O_2]_L)\theta^{T-20} \quad (4.45)$$

where $k_{ov,L,O2}a(20°C)$ is the overall oxygen transfer coefficient expressed at 20°C, $[S_{O2}]^*_{L,20}$ is the saturation concentration of oxygen at 20°C.

In comparison to clean water mass transfer, correction factors include several components:

- α is to take into account the modified resistance to transfer due to solutes and solids. The α-factor is further described above;
- F is a fouling factor defined as the ratio of the aeration system performance after use to new aeration system performance;

- β and τ are fundamentally incorporated via the activity correction (see Section 4.4.3) and the van't Hoff equation (equation (4.43)), respectively;
- Ω, the depth correction term, can be experimentally determined as proposed in ASCE (2007). Alternatively, the pressure impact may be taking into account considering an overpressure corresponding to half of the diffusers submergence (e.g. Baeten *et al.*, 2020) or to 33% of it (Gillot *et al.*, 2005). Ω, is in this case written as follows:

$$\Omega = \frac{\Delta P + P_{atm}}{P_{atm}} = \frac{\gamma h_{imm}(101.3 / 10.33) + P_{atm}}{P_{atm}} \tag{4.46}$$

where h_{imm} is the aerator submergence depth (m), assumed to be equal to liquid height above the aerator h, P_{atm} the atmospheric pressure (hPa), γ the correction factor due to the overpressure (0.33–0.5) (note: not activity).

The α-factor is a composite parameter that integrates the effect on oxygen transfer of a number of factors including the liquid matrix (surfactant concentrations, mixed liquor suspended solids concentrations, rheological properties, mean cell residence time and microbial activity) and the physical system characteristics (tank geometry and aeration system) (Amaral *et al.*, 2019; Gillot & Héduit, 2008). Moreover it has been reported to change both in time as in space (Amerlinck *et al.*, 2016), calling for implementing dynamic alpha factors (Jiang *et al.*, 2017). A dynamic (time dependent) α based on a pseudo conserved mass balance state is further presented in Bencsik *et al.* (2022). Multiple studies have been dedicated to the α-factor determination and the cause of its variability (to find out more about these impacts, references are listed in Amaral *et al.*, 2019).

4.6 CONCLUSIONS AND RECOMMENDATIONS

This chapter has developed the basis for gas-transfer theory, utilising principles of species activity from Chapter 2. It has been demonstrated that consistent principles of gas transfer are wholly integrated across the water cycle, including environmental gas transfer, aeration, anaerobic digestion, and emerging units such as stripping. Even in a moderate matrix such as mainline aeration, non-ideality and temperature dependency is critical (and incorporated in current practice), and this becomes even more so when dealing with concentrated matrixes, which exist in anaerobic digestion and ammonia stripping. The field of oxygen transfer from air is highly developed, and the standard practice identified above, while using different nomenclature and aggregated function forms, is wholly consistent with the more generalised approach to gas transfer developed in this chapter.

As noted in Section 4.2, two approaches can be taken to incorporate ion activity and non-ideality. Both these approaches can be taken in all circumstances of gas–liquid transfer, but some are more suitable in specific instances.

In the first, non-ideality is incorporated in determining saturation concentration, (using $[S_i]_L{}^*$). This is consistent with the oxygen transfer approach detailed above, and is the simplest implementation where species i does not form ion pairs. Also, the gas partial pressure p_i is generally a parameter (e.g., atmospheric partial pressure) and is therefore independent of liquid processes. We therefore recommend the saturation in this approach for non-pairing species when the transfer is gas → liquid.

When liquid → gas transfer dominates (e.g., strippers, anaerobic digesters), where there are multiple species forming, and when the species participate in pairing reactions (acid–base or other) the activity form $\{S_i\}_L$ has advantages, since p_i is commonly an output, and $\{S_i\}_L$ is explicitly calculated ($[S_i]_L$ would need to be re-calculated using the activity coefficient).

doi: 10.2166/9781780409832_0055

Chapter 5
Chemical oxidation and reduction

Damien J. Batstone[1], *Imre Takacs*[2] *and Xavier Flores-Alsina*[3]

[1]*Australian Centre for Water and Environmental Biotechnology, The University of Queensland, St Lucia, Brisbane, QLD 4072, Australia*
[2]*Dynamita, Roquesteron, Provence-Alpes-Côte d'Azur, France*
[3]*PROSYS Research Center, Department of Chemical and Biochemical Engineering, Technical University of Denmark, Building 229, DK-2800, Kgs. Lyngby, Denmark*

Chemical oxidation and reduction refer to spontaneous (not necessarily instantaneous) conversion of ions to a higher or lower oxidation state (Stumm & Morgan, 1996). Almost all spontaneous chemical redox reactions discussed here (e.g., iron reduction with sulfide as electron donor) is microbially catalysed (Kip & van Veen, 2015), but abiotic metal oxidation or reduction also exists as a spontaneous reaction, albeit slower. This chapter specifically excludes biochemically mediated reactions resulting in biomass production, which are covered extensively by biochemical models such as the ASM and ADM series (Batstone *et al.*, 2002a; Henze *et al.*, 2000). Biochemically mediated sulfur oxidation (by sulfur oxidising bacteria – SOB) and reduction (by sulfur reducing bacteria – SRB) is likewise out of scope (Hao *et al.*, 2014; Zhang *et al.*, 2021). The scope of this chapter is mainly metallic species oxidation with a relevant oxidant (e.g., O_2), or reductant (e.g., H_2, H_2S).

Chemical oxidation and reduction is a fundamental driver of chemistry in geological processes, often coupled with atmospheric biological sulfide oxidation (Torres *et al.*, 2014). In wastewater, metals are either in an elemental state or as ions in a stable oxidation state. While elemental metals can oxidise in a wastewater environment, the time scale of elemental metal is generally too slow to have a significant impact on wastewater process models except where the metal is deliberately oxidised via electrochemistry (e.g., in electrocoagulation (Moussa *et al.*, 2017)). Therefore, the key elements to consider are those which are found in wastewater (or added as oxidant/reductant), and oxidise/reduce between the oxidation reduction potential (ORP) range of +400 mV found in aerated tanks and the −500 mV found in anaerobic digesters (Li & Bishop, 2001; Zhu *et al.*, 2021). It is noted that chlorination disinfection systems and oxidation (ozone, advanced oxidation) can reach the 1 V+ range, where additional considerations may be needed (particularly where reduced metals are a significant sink of the oxidant).

Metal oxidation and reduction have a limited impact directly on biology, as the concentrations are too low to act as a significant source/sink of electrons (unlike sulfur). The main reason to include oxidation and reduction is to either (1) represent the different precipitation and speciation behaviour of different oxidation states, or (2) to simulate specifically transformation or removal of the metal as a contaminant. This chapter mainly focuses on scenario (1).

Metals which oxidise/reduce in the relevant range, and are present in wastewater are largely limited to line one transition metals (Ti through Zn), of which Fe and Mn are most relevant and available. Elements such as Ti are present, but have stable precipitates. Mn is most important in drinking water (where it is removed, because it forms a quality issue when oxidised from Mn^{2+} to Mn^{4+}) (Tobiason *et al.*, 2016). Fe oxidation and reduction is ubiquitous, and has the strongest effect across the engineered water cycle which is the focus here. Elements such as chromium or uranium may be deliberately reduced to less toxic forms [ref] through anaerobic treatment, but this has a limited impact.

5.1 ORP, $p\varepsilon$ AND POURBAIX DIAGRAMS

Oxidation–reduction reactions are coupled between compounds involving exchange of electrons. For example, the oxidation of ferrous ions to form ferric ions with oxygen has a net reaction (Stumm & Morgan 1996):

$$Fe^{2+} + \tfrac{1}{4}\,O_2 + H^+ \leftrightarrow Fe^{3+} + \tfrac{1}{2}\,H_2O \tag{5.1}$$

This net reaction is subject to equilibrium in activity (not concentration) as explained in Chapter 2, though it may not be instantaneous.

The net reaction can be represented by two hypothetical half-reactions, which explicitly include the electrons. Oxygen is reduced while ferrous is oxidised.

$$Fe^{2+} \leftrightarrow Fe^{3+} + e^- \quad \text{(oxidation)} \tag{5.2}$$

$$\tfrac{1}{4}\,O_2 + H^+ + e^- \leftrightarrow \tfrac{1}{2}\,H_2O \quad \text{(reduction)} \tag{5.3}$$

A higher activity of oxygen therefore represents a higher capability of the solution to donate electrons (i.e., oxidise by its reduction), while a higher activity of ferrous represents a higher capability of the solution to accept electrons (i.e., reduce by its oxidation).

Similarly to pH (the capacity of the solution to donate or accept protons), the electron activity, or $p\varepsilon$ can be defined, which is the relative tendency of a solution to accept or donate electrons (Stumm & Morgan 1996). Electrons do not actually exist as free electrons in solution, but represent the net capability and concentration of reduced species over oxidised species.

$$p\varepsilon = -\log\{e^-\} \tag{5.4}$$

Taking the equilibrium relationship for the half-reaction in equation (5.2)

$$\frac{\{Fe^{3+}\}\{e^-\}}{\{Fe^{2+}\}} = K_{\text{feox}} \tag{5.5}$$

where K_{feox} is the equilibrium constant for oxidation, can be related to free energy, and in this case is $\log(K_{\text{feox}}) = -13$. This is the reverse of the Fe reduction reaction ($\log(K_{\text{feox}}) = -13$), and the standard protocol is to represent the reduction reaction.

Considering equations (5.4) and (5.5), it can be seen that $p\varepsilon$ can be determined as

$$p\varepsilon = -\log K_{\text{feox}} + \log\frac{\{Fe^{3+}\}}{\{Fe^{2+}\}} \tag{5.6}$$

Log K values for a range of reduction half-reactions is given in Stumm and Morgan (1996).

The $p\varepsilon$ can be expressed in voltage (solution $E_{H\text{-}}$ ORP) by the Nernst equation

$$E_H = \frac{2.3RT}{F}p\varepsilon \tag{5.7}$$

where F is the Faraday constant (96 485C mole^{-1}). At 25°C, 2.3RT/F~0.059 V mole^{-1}. Likewise, the standard half-cell electrode potential for a redox pair (normalised to a single electron) can be calculated from the $p\varepsilon^0$ (log K_{red}).

$$E_H^0 = \frac{2.3RT}{F} p\varepsilon^0 \tag{5.8}$$

For Fe^{3+}/Fe^{2+}, this is +0.77 V, H^+/H_2, 0 V, $O_2 + H^+/H_2O$, +1.23 V (Stumm & Morgan, 1996). Pairs with a higher redox potential (in oxidised form) oxidise compounds with a lower redox potential (in reduced form). Therefore, Fe^{3+} will reduce with hydrogen to form Fe^{2+} and Fe^{2+} will oxidise with oxygen to form Fe^{3+}.

Naturally, the activity of a compound is impacted by ion pairing as discussed in Chapter 2. The most important pairing is acid–base pairing, which is why redox is strongly linked to pH. At low pH, Fe^{2+} and Fe^{3+} are both fully ionised, but as pH rises, both progressively and asymmetrically form iron hydroxides. This is used to generate Pourbaix (or stability) diagrams, which indicate the stable form of a redox component according to pH and ORP. Other compounds may also be included (e.g., bicarbonate in this case). As can be seen in Figure 5.1, at low pH, the transition between reduced and oxidised Fe is 0.77 V, but as pH increases, the ORP region for Fe^{2+} decreases and narrows. In fact, at pH 8, the redox ORP is 0 V.

ORP is measured using an inert Pt electrode against a reference electrode (generally Ag/AgCl). This is an ambiguous measurement due to varying reactivity (with the solution) of different compounds

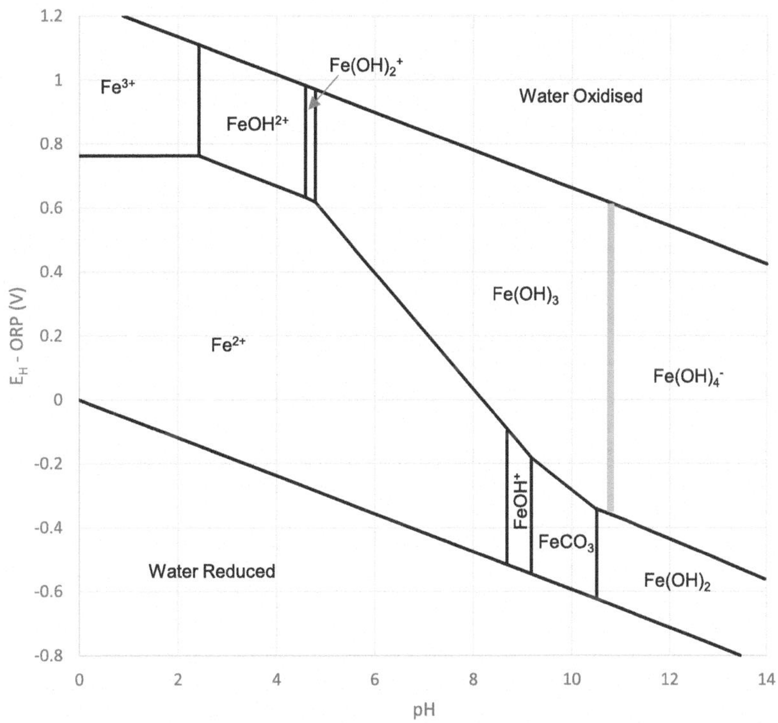

Figure 5.1 Pourbaix (stability) diagram for 0.01 ppm iron with 100 ppm HCO_3. (*Source*: data from Hem (1961)) $Fe(OH)_4^-$ formation point is approximate at this concentration.

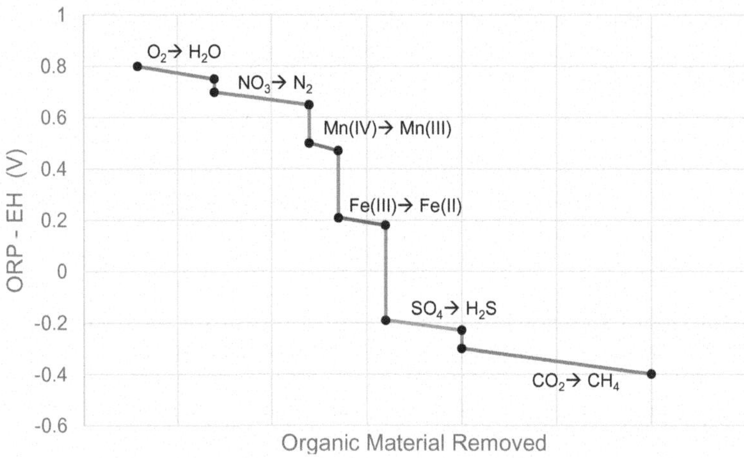

Figure 5.2 Redox ladder, showing ORP (E_H) progression as oxidants are consumed. (*Source*: Data from Appelo & Postma (2005), Stumm & Morgan (1996), at pH 7). The actual transition points and shape depend on the concentration of oxidising and reducing species present, as well as other factors such as temperature.

and modification of the electrode with solution compounds, and various factors discussed below relating to solution at non-equilibrium state (Appelo & Postma, 2005).

ORP can in theory (for a solution at equilibrium, which is almost never the case) be calculated from the ratio of oxidised:reduced compound, estimated from a model using equations (5.6) and (5.7). However, equation (5.6) is on a log scale, and reactants are often fractional (e.g., equation (5.3)), which means that a 10× change in for example, DO only results in a minor change in ORP. pH also has a large impact (as seen in equation (5.1)). There are substantial transitions in ORP depending on whether the system is dominated by oxygen reduction (aerobic), nitrate reduction (anoxic), nitrite reduction (anoxic), sulfate reduction (anaerobic) or methanogenesis (anaerobic). This results in the redox ladder, as shown in various forms (see also Figure 9.1 in Appelo & Postma (2005) and Figure 8.14 in Stumm & Morgan (1996)) (Figure 5.2). This is not a strict ladder, as redox reduces along the rung as oxidant is consumed, and the oxidising agent switches between various forms of the principal compound (e.g., SO_4, S^0, polysulfide etc).

ORP is difficult to model as an objective output, and cannot be reliably measured (as an objective output). While (as discussed above), certain environments are likely to have a given ORP, there is almost no correlation between observed and pair-calculated ORP (Lindberg & Runnells, 1984). If ORP is needed as a model output, it can be determined empirically from the key states of DO, nitrate (and/or ammonia), nitrite, sulfate, and methane (Varga *et al.*, 2018), noting a log scale is preferred, though pH should also be a key factor. Hydrogen and sulfide are highly effective, due to reactivity with ORP probes, but are only present at very low ORP.

To simulate iron oxidation/reduction, this is not required, since iron will readily oxidise in aerobic (0.5–0.75 mV) or anoxic (0.1–0.6 mV) environments, and reduce in anaerobic environments (<−0.1 mV). Manganese is similar, oxidising when oxygen is present, and reducing in anoxic zones at very low nitrate or nitrite.

While metal ion oxidation and reduction are explained from the geochemical (and thermodynamic perspective) here, there is limited difference to biochemical conversion of other species, except that they are autocatalytic (i.e., depend on the concentration of biomass, and result in biomass as a product). Metal reduction can be handled in a way similar to how biochemical processes are – that is, inhibited by oxygen, or simply omitted in the relevant process.

5.2 RECOMMENDED APPROACH

As noted above, implementation of metal species oxidation can be relatively simple, if it is known that a metal oxidises readily under aerobic conditions and reduces readily under anaerobic conditions. In these cases, a simple second-order kinetic can be used with utilises relevant electron sources and sinks. Omitting half-saturation kinetics, and choosing a high second-order coefficient will ensure that the process outcompetes biochemical reactions ((Flores-Alsina *et al.* 2016) suggest $10^9\,d^{-1}$). The approach of Flores-Alsina *et al.* (2016) is simplified here, with electron flowing to form oxygen or nitrate under aerobic or anoxic conditions, and taken from sulfide or hydrogen under anaerobic conditions. The metal can be expressed in COD units (reduced version having positive COD), but it is generally more common to represent metallic species as mgMe/L (as is done here). If sulfide is the reductant, it is common to consider elemental sulfur S^0 as the product of sulfide oxidation (Hauduc *et al.* 2019) since subsequent oxidation of elemental sulfur to sulfate is relatively slow, and may not be thermodynamically favourable. As for metal species, sulfur may be expressed as element concentration or COD units, with sulfate being the 0-COD state (element is taken here). The availability of ferric and ferrous ions will impact on the formation of HFO (see chapter 3), and indeed, ferric will rapidly complex to form HFO when oxidised from ferrous (Figure 5.2, Chapter 3), which is why HFO (X_{HFO}) is included instead of $S_{Fe(III)}$ in Table 5.1. This implements direct reduction of HFO under anaerobic conditions, which will then directly induce phosphorous release from HFO.

This is effective for the majority of the metals being considered, under aerobic and anaerobic conditions, but alternatives may need to be considered where metal oxidises or reduces under partially oxidising or reducing conditions.

Table 5.1 Stoichiometry and kinetics for iron oxidation and reduction.

Process (*j*) ↓ Component (*i*) →	S_{O2}*	S_{NO3}*	S_{N2}*	S_{h2}**	S_{IS}	X_{S0}	$S_{Fe(II)}$	X_{HFO}	ρ_j
Oxidation of $S_{Fe(II)}$ to $S_{Fe(III)}$ on S_{O2}*	−0.1432						−1	1	$k_{Fe2Fe3}S_{Fe(II)}S_{O2}$
Oxidation of $S_{Fe(II)}$ to $S_{Fe(III)}$ on S_{NO3}*		−0.0501	−0.0501				−1	1	$k_{Fe2Fe3}S_{Fe(II)}S_{NO3}$
Iron (III) reduction (H_2)**				−0.1432			1	−1	$k_{Fe3Fe2}X_{HFO}S_{H2}$
Iron (III) reduction (H_2S)**					−0.2879	0.2879	1	−1	$k_{Fe3Fe2}X_{HFO}S_{IS}$
Units	Oxygen (g O_2.m^{-3})	Nitrate + Nitrite (g N.m^{-3})	Dinitrogen (g N.m^{-3})	Hydrogen (g COD.m^{-3})	Hydrogen sulfide (g S.m^{-3})	Elemental Sulfur (g S.m^{-3})	Ferrous iron (g Fe.m^{-3})	HFO (g Fe.m^{-3})	

* *Aerobic/Anoxic conditions*
***Anaerobic conditions*

5.3 CONCLUSIONS

While these reactions are biocatalysed (and result in marginal biomass yields), they also occur under abiotic conditions, and are generally considered in models as abiotic reactions, included as above. Inclusion of metal oxidation and reduction in plant-wide models is a recent addition to the modelling area, but as with most aquatic chemistry, the fundamentals are well understood.

This chapter focuses on processes relevant to plant-wide biochemical modelling, and does not address chlorination, which is an interesting and relevant chemical oxidation process (chlorine decay is reduction of free chlorine with organics and reduced metals), and since hypochlorous acid is a weak acid (and forms chlorine gas), with the form (acid or base) influencing pathogen destruction effect, decay, volatilisation, and other factors such as reaction with ammonia to form chloramines (Tchobanoglous *et al.*, 2003). This extends to environmental decay kinetics, essential to limiting environmental impact of free chlorine. While pH is used to assess chlorination effect, it is included as an effect rather than an integrated model component, and particularly for decay, inclusion is largely empirical (Khawaga *et al.*, 2018; Monteiro *et al.*, 2014). The use of an advanced speciation model (including gas transfer) has the potential to enhance chlorine addition and transformation, even if the underlying decay and pathogen destruction model remains a simple first-order process (Tchobanoglous *et al.*, 2003).

doi: 10.2166/9781780409832_0061

Chapter 6

Implementation in plant-wide models

Xavier Flores-Alsina[1], Kimberly Solon[2], Hannah Feldman[3], Christian Kazadi-Mbamba[4], Tamara Fernández-Arévalo[5], Paloma Grau[6], Ulf Jeppsson[7], David Ikumi[8], Krist V. Gernaey[1] and Damien J. Batstone[4]

[1]PROSYS Research Center, Department of Chemical and Biochemical Engineering, Technical University of Denmark, Building 229, DK-2800, Kgs. Lyngby, Denmark
[2]Department of Green Chemistry and Technology, Ghent University, Coupure links 653, B-9000, Gent
[3]Novozymes A/S, Hallas Alle 1, DK-4400, Kalundborg, Denmark
[4]Australian Centre for Water and Environmental Biotechnology, The University of Queensland, St Lucia, Brisbane, QLD 4072, Australia
[5]CEIT-Basque Research and Technology Alliance (BRTA), Manuel Lardizabal 15, 20018 Donostia/San Sebastián
[6]University of Navarra, Manuel Lardizabal 15, 20018 Donostia/San Sebastián
[7]Lund University, Division of Industrial Electrical Engineering and Automation, Department of Biomedical Engineering, Lund University, Box 118, SE-221 00, Lund, Sweden
[8]Water Research Group, Department of Civil Engineering, University of Cape Town, Rondebosch, 7700, South Africa

6.1 INTRODUCTION

This chapter provides a description of how the physicochemical model (PCM) framework developed in this scientific and technical report (STR) (ion speciation and pairing, precipitation/dissolution, gas transfer and oxidation/reduction) can be applied to conduct integrated assessments. The PCM is linked to standard models used in wastewater engineering for example ASM1, 2d, 3 (Henze *et al.*, 2000) and ADM1 (Batstone *et al.*, 2002a). In other cases, special plant-wide biochemical models (PWBM) are developed for this purpose. In the first case study, the chemical oxygen demand (COD), N, P, S and Fe fluxes are studied with the benchmark simulation model No 2 (BSM2) plant layout. Next, a list of full-scale applications is presented where models have been used to answer different questions: chemical dosage, suitability of streams, potential phosphorus recovery and/or aeration efficiency. The chapter uses the general (model-based) notation of X_i for particulate concentration states, S_i for soluble concentration states, and Z_i for implicit algebraic states. Note that other unit and plant-wide implementations may have additional state concentration fractions (e.g., colloidal C_i – (Hauduc *et al.*, 2018)), but in most of the cases in this chapter, the colloidal fraction is lumped with particulates (X_i).

6.2 CASE STUDY #1: IMPLEMENTATION IN THE BSM2 AND PERFORMANCE ANALYSIS

This section describes the impact of operational strategies on N, P, S fluxes in the BSM2 plant layout (see Figure 6.1). The flow diagram consists of a primary clarifier (PRIM), an activated sludge unit (ASU), a secondary settler (SEC2), a sludge thickener (THK), an anaerobic digester (AD), a storage tank (ST) and a dewatering unit (DEW). The main modification with respect to the original design (Gernaey *et al.*, 2014) is in the ASU. An anaerobic section without oxygen (O_2) and nitrates (NO_x) is needed to promote anaerobic phosphorus release and to provide the phosphorus accumulating organisms (PAOs) with a competitive advantage over other bacteria. Phosphorus release from the breakdown of polyphosphates (PO_4^{3-}) provides the energy required for anaerobic uptake of polyhydroxyalkanoates (PHA). Next, PAOs grow using intracellular storage products (i.e., PP) as a substrate while taking up N and P as nutrients in the anoxic and aerobic reactors with NO_x (with less efficiency) or O_2 as electron acceptors, respectively (Figure 6.1). Additional details on the design and default operational conditions can be found in Gernaey *et al.* (2014). It is important to highlight that this configuration does not represent an optimal design to remove P, because the biological P removal is dependent on the N removal *via* the nitrate load recycled to the anaerobic reactor *via* the underflow recycle (i.e., excess nitrate may cause the anaerobic reactors to become anoxic). Nevertheless, it exemplifies the retrofit of many (C, N removal) plants adapting their plant layout to satisfy new and stricter effluent requirements.

There are two core models predicting the behaviour of the AS and the AD units. A modified version of the Activated Sludge Model No. 2d (ASM2d) is selected to describe organic carbon, nitrogen and phosphorus transformations in the biological reactor (Henze *et al.*, 2000). The Anaerobic Digestion Model No. 1 (ADM1) version, implemented in the plant-wide context provided by the Benchmark Simulation Model No. 2 (BSM2) (Batstone *et al.*, 2002a; Rosen *et al.*, 2006) is extended with P, S and Fe interactions (Flores-Alsina *et al.*, 2016). These two models are linked with the PCM described in the previous chapters. Additional information about model structures can be found in Gernaey *et al.* (2014) and Solon *et al.* (2017).

Figure 6.2a shows the mass balances for the default operating conditions. Model predictions show that 40, 39 and 30% of the incoming COD, C and N are stripped to the atmosphere after biological treatment in the AS. More specifically, 1, 24 and 42%, respectively, leave the plant *via* the effluent and

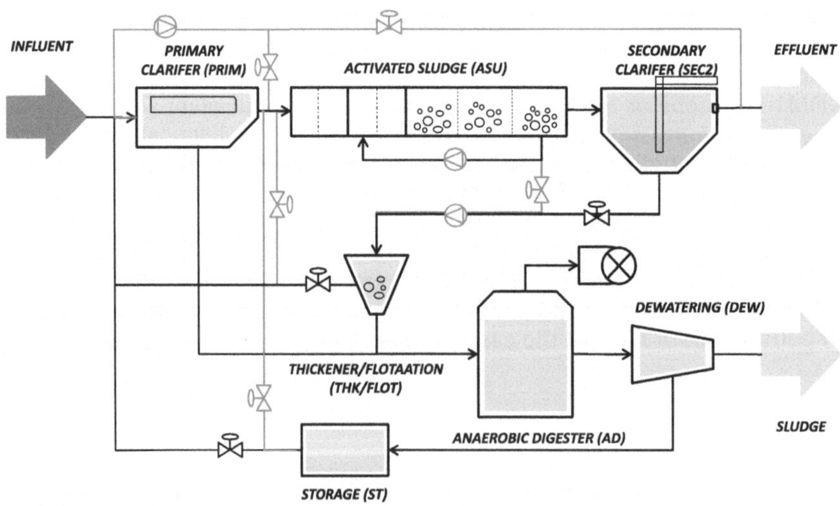

Figure 6.1 Schematic diagram of the BSM2 plant.

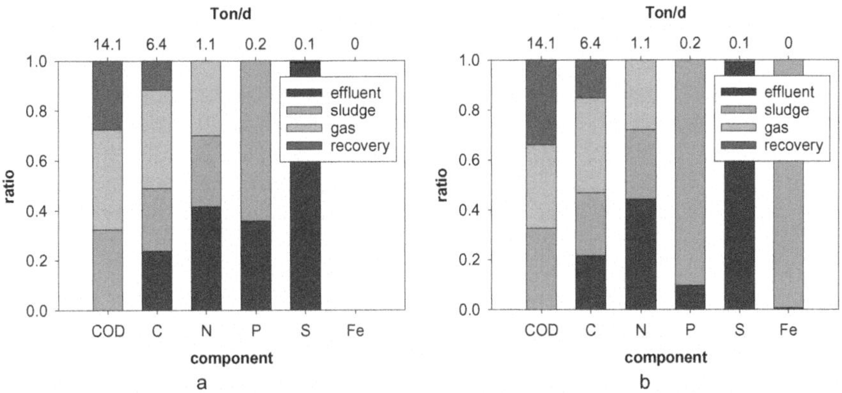

Figure 6.2 Mass balances for COD, C, N, P, S and Fe for the default operational conditions (a) and after adding Fe (b).

32, 25 and 28% end up in the sludge to be disposed. It is important to highlight that 29 and 12% of the COD and C are transformed in the AD into methane (CH_4) (and potentially recovered as energy). Almost all the incoming S (99%) leaves the plant dissolved in the effluent stream. Only a small fraction (1%) is stripped in the AS, transformed in the AD into sulfides (H_xS) and potentially recovered as sulfur mineral (S_0) or accumulated in the sludge. With respect to incoming P, around 64% goes to the sludge line. The remaining 36% leave the plant *via* the effluent. As an alternative scenario, Figure 6.2b shows the mass balances when Fe is added. The main differences rely on the quantity of recovered COD and C (37 and 15%) in the AD as a result from: (1) chemical enhanced primary clarification; and (2) a lower degree of H_2S inhibition due to iron sulfide (FeS) precipitation. The other significant difference depends on the formation of Fe–P compounds, which substantially increases P removal efficiency.

Figure 6.3 provides additional details about the N, P, S and Fe transformations at different locations within the BSM2 plant and for the two evaluated operational strategies. When it comes to N (see Figure 6.3a), 50% arrive as ammonium (NH_x), while the remaining fraction is associated with the organic solubles ($OrgN_{sol}$) and particulates ($OrgN_{part}$) (see Influent). After biological treatment, the overflow of the secondary settler (Effluent) is mainly comprised of ammonium (NH_x) or nitrates (NO_X) (70%). Simulated (N) effluent values ($TKN = 2.9$ g N m^{-3} and $TN = 9.1$ g N m^{-3}) are well below the limits set by the BSM evaluation limit ($TKN_{limit} = 4$ g N m^{-3} and $TN_{limit} = 15$ g N m^{-3}). The N load going to the sludge line is associated with $OrgN_{part}$ (99%) (AD_{in}). During AD (AD_{out}), a substantial amount of the OrgN is hydrolysed to NH_x (50%). The remaining particulate N fraction is retained after dewatering. As a result, N in the sludge (DW_{over}) is mainly associated with $OrgN_{part}$ (80%). A minor part is also retained as struvite in the inorganic particulate fraction ($InorgN_{part}$). In the reject water stream (DW_{under}), 99% of the N is ammonium (NH_x). No substantial differences can be observed between the two evaluated strategies (with and without adding Fe) (see Figure 6.3b).

In the influent, there is a 35–40% distribution between phosphates (PO_4^{3-}) and organic solubles ($OrgP_{sol}$) vs. particulates ($OrgP_{part}$) (see Figure 6.3c). After treatment, most of the P leaving the plant (effluent) is in the form of H_xPO_4 (68%). The predicted effluent value for TP is 3.5 g P m^{-3}, which for this particular case, is above the defined regulatory limits ($TP_{limit} = 2$ g N m^{-3}). At AD_{in} most of the P captured by PAOs is part of polyphosphates (PP) (67%) and $OrgP_{part}$ (32%). After digestion (AD_{out}), all PP decay and most of the organics have hydrolysed in the AD effluent, P can be found as H_xPO_4 (62%), bound with calcium (Ca) and magnesium (Mg) ($InorgP1_{part} = 26\%$) and either $OrgP_{sol}$ or $OrgP_{part}$ (2%). $InorgP1_{part}$ and $OrgP_{part}$ will end up being part of DW_{over} (95%), while phosphates (PO_4^{3-}) and $OrgP_{sol}$ (99%) return to the AS unit *via* the reject water stream (DW_{under}). The addition of Fe has a strong

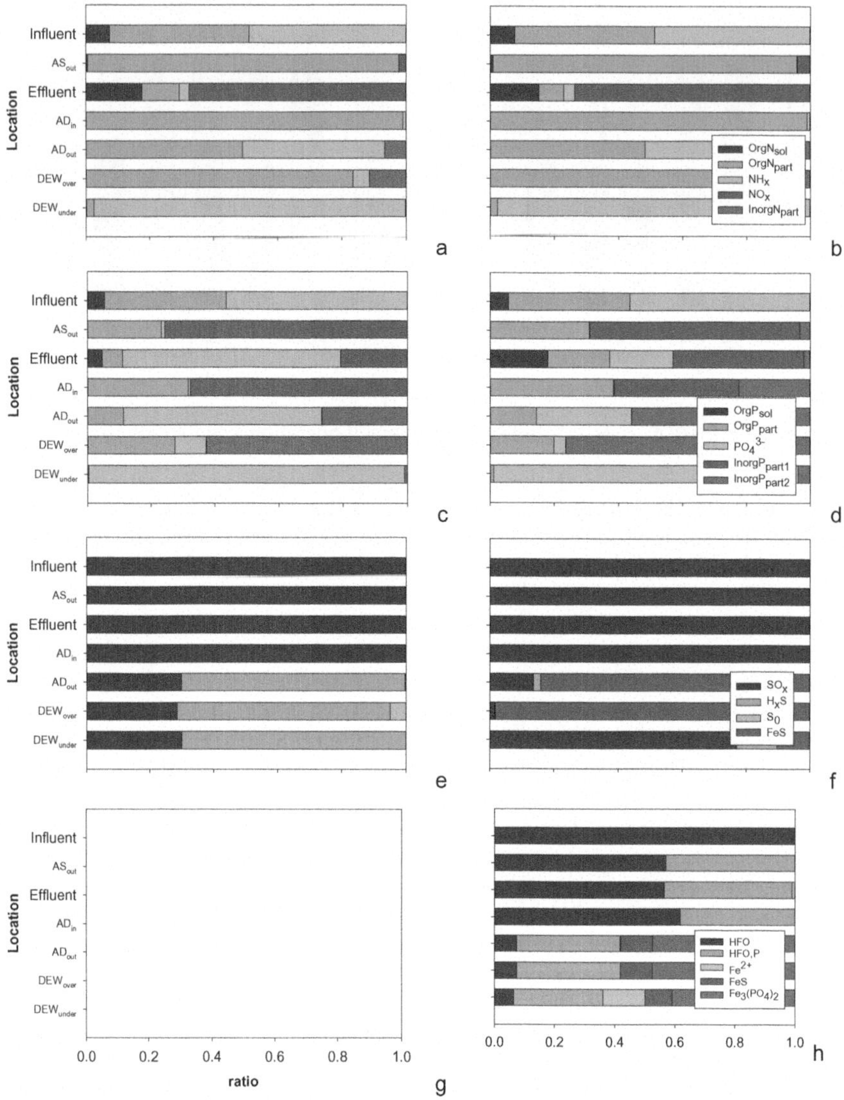

Figure 6.3 N (a, b), P (c, d), S (e, f) and Fe (g, h) distributions at different plant locations. Figures on the left (a, c, e, g) represent default operational conditions while figures on the right (b, d, f, h) represent scenario with Fe addition. Note that the figure g is empty because there is no Fe addition.

impact on the P fluxes within the plant. As mentioned before, first and the foremost, the efficiency of P removal is substantially increased (TP = 0.96 g m^{-3}) (see Figure 6.2). Second, the distribution of the P forms through the plant is altered, particularly at the output of the AS and the inlet/outlet of the AD. More specifically for the first two, inorganic iron precipitates ($InorgP2_{part}$) become almost as important as $OrgP_{part}$ and PP. It also affects the distribution of PO_4^{3-} and $InorgP1_{part}$ since new compounds are being formed, that is, $Fe_3(PO_4)_2$.

Figure 6.3e reveals that all the incoming S is assumed to be sulfates (SO_x). There is a minor reduction to sulfides (H_xS) in ANAER, but this is rapidly re-oxidized or stripped to the atmosphere in AER1, 2 and 3. As a result, the output of the AS and the effluent is 100% sulfates (SO_x) (see AS_{out} and Effluent). The main transformation of S takes place in the AD, where the incoming S arriving via primary and secondary sludge flows is reduced to sulfides (H_xS) again. This changes the S distribution at AD_{out} to 70% H_xS and 30% SO_x. A part of this S also goes to the gas phase and is removed from the liquid stream (see Figure 6.2). Since both are soluble compounds, they have the same distribution after dewatering (see DW_{over} and DW_{under}). The addition of Fe also changes the entire S dynamics (see Figure 6.3f). More specifically, at AD_{out}, most of the S (80%) is captured as iron sulfide (FeS). This also affects the quantity of S stripped to the gas phase and the COD conversion to CH_4 (see Figure 6.2). The formation of iron sulfide (FeS) and the S concentration in particulate form also causes differences due to dewatering. In the sludge cake (DW_{over}), S is mainly FeS (98%). In the reject water (DW_{under}), SO_x is the dominant species (80%).

When Fe is added it is assumed to be automatically converted into hydrous ferric oxides (HFO) (see Figure 6.3h). Those react (absorbed, co-precipitated) with free phosphates PO_4^{3-} and form HFO, P. A part of these HFO can be reduced to ferrous iron (Fe^{2+}) in ANAER1 and 2, but as it happened with S, this recently formed Fe^{2+} is re-oxidized again to HFO in AER1, 2 and 3. Both HFO and HFO, P are sent to the AD (60 and 40% distribution). In the AD, HFO, P can dissolve in HFO. This pool of HFO can be reduced to Fe^{2+} and then precipitate with S and form iron sulfide (FeS) or with P and form vivianite ($Fe_3(PO_4)_2$). The latter two species account for almost 50% of all the Fe species. As mentioned before, the formation of $Fe_3(PO_4)_2$ strongly affects $InorgP1_{part}$ (see Figure 6.3b). The distribution of Fe in the dewatering overflow is the following: soluble forms are mainly in the reject water line, while all the particulate forms (HFO, HFO, P, FeS and $Fe_3(PO_4)_2$) are mainly found in the waste sludge.

As is evident from the analysis of this plant-wide system, the impact on the nitrogen cycle is limited but significant. However, the iron–phosphorus–sulfur interactions are tightly linked, and it strongly recommends describing phosphorus dynamics with inclusion of iron and, which requires additional oxidation/reduction, as well as a complex PCM.

6.3 CASE STUDY #2: EFFECT OF PH AND FE/P RATIO ON PHOSPHORUS REMOVAL IN AN MBR PILOT

The second case study presents the results of applying the developed PCM when assessing the impact of the Fe/P molar ratios on effluent phosphates (PO_4^{3-}) concentrations resulting from different dosage strategies. A series of simulations are conducted at different pH values (Kazadi Mbamba *et al.*, 2019). The plant layout is based on a membrane biological reactor (MBR) pilot plant under study is located at the R&D facility Hammarby Sjöstadsverk, which is adjacent to Henriksdal wastewater treatment plant (WWTP) (850 000PE), the largest in Stockholm (Sweden). The pilot-plant treatment process consists of a conventional primary treatment including a pre-aeration step, where iron sulfate ($FeSO_4$) is added, followed by seven bioreactors operated as a four-stage modified Ludzack-Ettinger (MLE) process consisting of two anoxic (BR1/ANOX and BR2/ANOX), one aerobic/anoxic (BR3/FLEX), two aerobic (BR4/AER and BR5/AER), one de-oxygenation (BR6/DEOX) and one post-anoxic (BR7/ANOX) (see Figure 6.4).

Similarly to case study 1, the core biological models are the modified version of ASM2d and ADM1 (Solon *et al.*, 2017).

The calibrated plant-wide model of the pilot study is used to study the impact of pH and molar ratio of Fe/P on the removal efficiency of phosphorus, which involves the precipitation of iron complexes, co-precipitation and adsorption of P (see Figure 6.4). As indicated in Figure 6.5, the model shows that under the three molar ratios (1, 1.5 and 2), decreasing pH in the bioreactors led to lower P concentrations in the effluent. For the lowest ratio (Fe/P = 1), the effluent P dropped from 0.4 gP m^{-3}

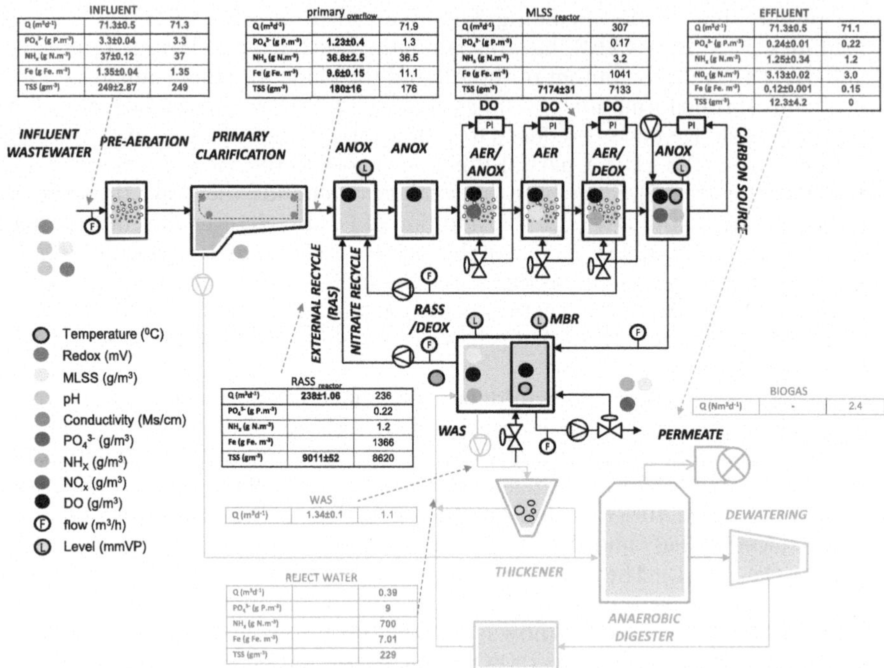

Figure 6.4 Schematic diagram of case study #2 with plant measurements, model predictions and sensor locations.

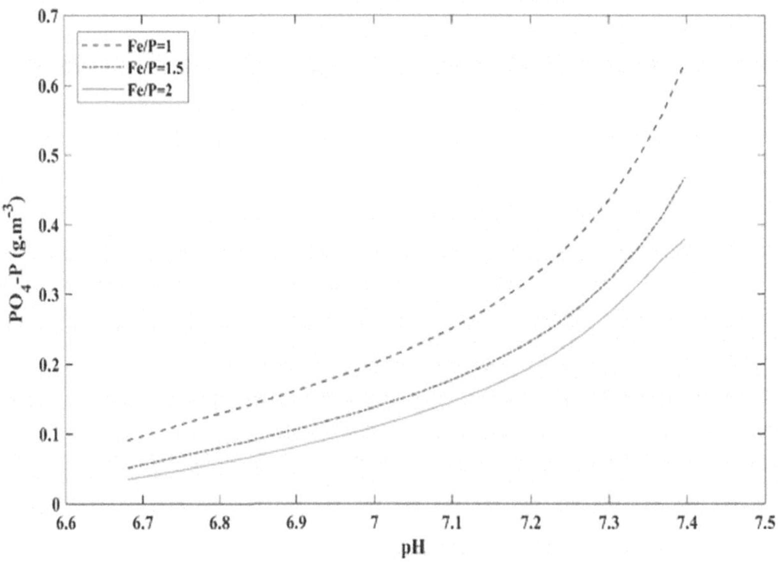

Figure 6.5 Impact of pH and Fe/P molar ratio on the effluent dissolved phosphorus concentration. (*Source*: Kazadi Mbamba *et al.*, 2019).

at higher pH (7.4) to 0.002 gP m^{-3} at pH 6.7. However, the relative differences of P concentrations in the effluent between the highest and lowest Fe/P ratios were higher (0.63) at higher pH but became smaller (0.009) at lower pH. This shows that the pH and the molar ratio of iron to phosphorus affect the competition between hydroxyl ions and phosphates for ferric ions at the point of addition. As a rule, the pH in aqueous environments affects the chemical distribution of the weak acid–base equilibria for systems such as phosphate and ferric iron complexation, which in turn control chemical precipitation and adsorption. The model results also demonstrate that increasing the concentration of iron salt (e.g., $FeSO_4$ and $FeCl_3$) in the MBR system may lead to a marginal increment on P removal. This modelling analysis shows the importance of using a robust PCM to capture the mechanisms of iron precipitation and provides detailed insight into the interactions between iron species and phosphorus within a biological nutrient removal with membrane separation.

6.4 CASE STUDY #3: IMPLEMENTATION IN A PWM AND RECOVERY OF STRUVITE

This section presents a study carried out at the Sur WWTP (Madrid) to evaluate optimal phosphorus management strategies based on the plant-wide modelling (PWM) methodology (Lizarralde *et al.*, 2019). The objective of the study is to carry out a model-based scenario analysis to evaluate the maximum quantity of struvite that can be recovered in the WWTP, as well as the analysis of the benefits on the plant operation. To that end, four different scenarios were compared in terms of struvite production, biogas production, sludge production, ferric chloride dosage, and pipeline blockage due to uncontrolled struvite precipitation. The plant layout consists of a pre-treatment, a primary clarifier and an activated sludge treatment for organic matter and phosphorus removal in the water line, and different sludge thickening processes, anaerobic digestion and a dehydration step with centrifuge in the sludge line. The biological treatment treats 260 000 m^3/d (2 937 000PE), is divided into six identical lines, and each of them contains an anaerobic zone, a facultative zone and an aerobic zone to allow the alternation of volumes of the anaerobic/aerobic phases and enable optimal activity of the PAOs.

The reference scenario (scenario A) is based on the current plant. In the second scenario (scenario B), a precipitation unit is added after the DEW to recover struvite from the digester supernatant. In the third scenario (scenario C), a redissolution tank is added to treat the secondary sludge in order to avoid pipeline blockage due to uncontrolled precipitation of struvite; and in the last scenario (scenario D), a fraction of the primary sludge is taken to the redissolution tank as a source of organic matter. More information on these configurations can be found in Lizarralde *et al.* (2019). Figure 6.6 shows a schematic of the plant layout, and the complete phosphate flows in the four scenarios.

The mathematical model used for this study is based on the PWM methodology proposed by Ceit-BRTA (Fernández-Arévalo *et al.*, 2014; Grau *et al.*, 2007; Lizarralde *et al.*, 2015). This methodology allows the straightforward construction of compatible mathematical models, ensuring elemental mass and charge continuity for any possible combination of unit-process models. Given the characteristics of the plant, the CNPprec_AnD category of the Ceit PWM library (Fernández-Arévalo *et al.*, 2017) was selected to reproduce the behaviour of the plant. The biochemical reactions considered in the model were the ones that are necessary to describe biological organic matter, P and N removal under different environmental conditions (aerobic, anoxic and anaerobic). The chemical transformations considered in the model were the weak acid–base and complex ion-pairing equilibrium reactions between volatile fatty acids (VFAs), inorganic carbon, N, P, Ca, Mg, K, Fe^{3+}, Fe^{2+}, Na and Cl. To describe the liquid–gas processes, six gaseous components and the mass exchange between the gaseous and aqueous phases were included: H_2O, O_2, CO_2, NH_3, H_2 and CH_4. Regarding the precipitation–redissolution reactions, the most relevant sparingly soluble salts proposed by Musvoto *et al.* (2000a) and their precipitation according to the equation proposed by Lizarralde *et al.* (2015) were considered, that is $CaCO_3$, $MgCO_3$, $Ca_3(PO_4)_2$, struvite, kstruvite and newberyite. Biochemical and physicochemical transformations were described using ordinary differential equations (ODE), while chemical reactions were described

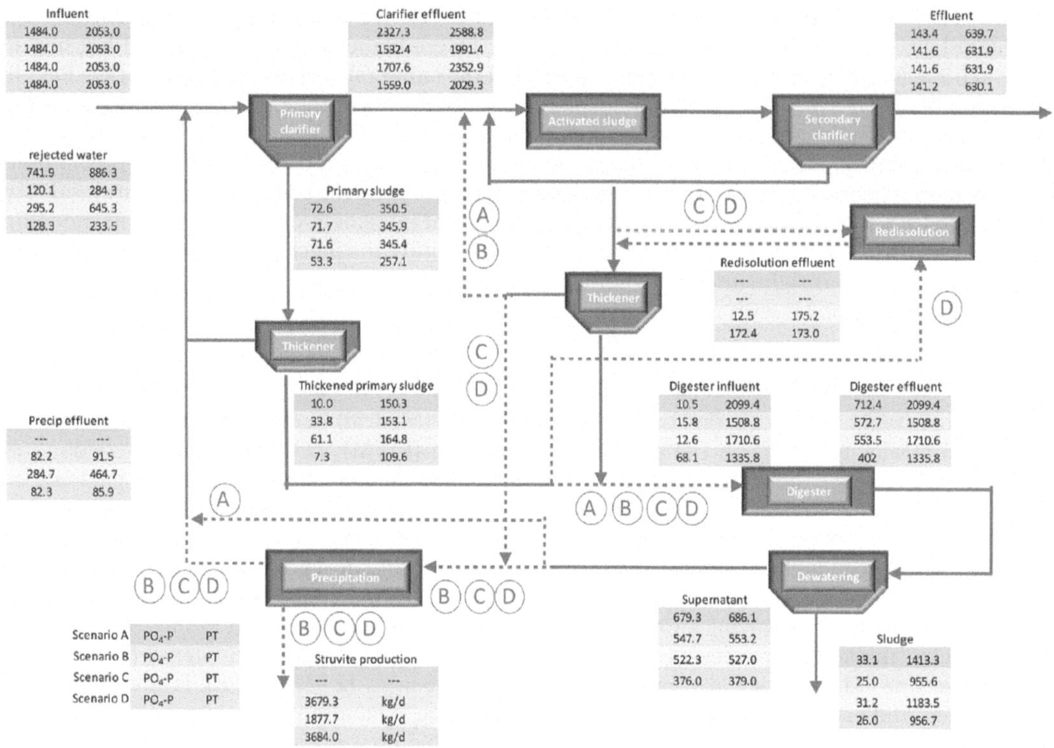

Figure 6.6 PO_4^{3-} mapping throughout the plant in the different scenarios. (*Source*: Lizarralde *et al.*, 2019).

using algebraic equations according to the Tableau method as described in Chapter 2 (Morel & Hering, 1993). Particularly acid–base and gas-transfer models are mathematically compatible with Chapters 2 and 3, respectively, and the precipitation model, is functionally similar (i.e., consideration of precipitation equilibrium with a rate expression).

Based on one year of operation, in the three new scenarios proposed in the study, struvite is recovered (Table 6.1). Based on the analysis by simulation, in scenarios B and D, the amount of PO_4^{3-}-P in the rejected water is reduced by around 80%, with a smaller reduction in scenario C (43%) due to a lower recovery of struvite. Although the recovery of struvite is similar in scenarios B and D, in scenario D the release of phosphorus occurs jointly in the redissolution tank (172.4 kg P/d) and in

Table 6.1 Comparison of performance of the different scenarios.

Variable	Scenario A	Scenario B	Scenario C	Scenario D
Struvite production (Tn/year)	–	1343	685	1345
$MgCl_2$ dosage (Tn/year)	–	684	250	691
NaOH dosage (Tn/year)	–	403	26	444
Uncontrolled struvite precipitation (Tn/year)	126	95	119	38
Sludge production (Tn/year)	19 070	17 809	18 821	17 829
$FeCl_3$ dosage (Tn/year)	4979	2490	4914	2733

the AD (402 kg P/day), reducing the release of phosphorus in the digester and therefore minimizing the risk of uncontrolled phosphorus precipitation. In scenario C, the amount of phosphorus released in the redissolution tank is very low. Mixing this diluted flux with the digester supernatant generates a very diluted influent to the precipitation unit, thus reducing the efficiency of the precipitation unit.

The magnesium chloride dosage, to provide with the required Mg^{2+} to form struvite, is directly dependent on the struvite recovered, and consequently higher magnesium chloride is required in scenario D (1.05%, Table 6.1). Sodium hydroxide, in turn, is associated with the pH of the streams. The amount of NaOH required in scenario D is theoretically 10.2% higher than in scenario B, since the flow from the redissolution tank is lower (6.9) than the pH in the effluent of the AD (7.4) in scenario D. These effects show the need to ensure a good aquatic chemistry and pH prediction throughout the plant.

In scenario B, the amount of uncontrolled struvite precipitation is 24.6% lower compared to scenario A. Controlled precipitation of phosphorus allows extracting phosphorus from the system and therefore reduces the amount of phosphorus released in the digester by hydrolysis or by redissolution. Even so, the key to minimizing uncontrolled struvite precipitation is to decouple phosphorus release from ammonium release. For this, redissolution prior to digestion is a good solution, if the reactor has sufficient organic matter for it. As can be seen in Table 6.1, controlled precipitation will allow us to reduce sludge production, as well as the need for $FeCl_3$.

To perform a plant optimization study or a feasibility analysis of incorporating a new technology, it is important that the model of the entire plant includes the liquid line, the sludge line and the interactions between them. When the biological phosphorus removal process is improved, a greater amount of phosphorus accumulates in the PAOs. This leads to a greater release of phosphates (PO_4^{3-}) in the anaerobic digestion process, which increases the uncontrolled precipitation of struvite. Hence, the importance of considering the overall plant configuration during a P management study. Another important aspect to consider is the pH, especially in the sludge line. The degree of struvite precipitation is a function of Mg^{2+}, NH_x, and PO_4^{3-} concentrations, but is strongly conditioned by the temperature and pH of the sludge. The activities of NH_x and PO_4^{3-} ions are strongly influenced by the pH of the medium, which makes pH a key factor for crystallization. Linked to this, it is important to consider the elemental mass continuity in the model, especially COD, Mg^{2+}, K^+, N, and P.

6.5 CASE STUDY #4: OXYGEN TRANSFER EFFICIENCY STUDY AND OPTIMIZATION USING A PWM

The fourth case describes a model-based study of the aeration process in Galindo WWTP (Bilbao, Spain) in order to: (1) assess the behaviour of the diffusion system by the measurement of the oxygen transfer efficiency (OTE) and (2) explore by simulation the effect of various design and operational parameters such as reactor height, bubble size, total suspended solids concentration or temperature over OTE. Galindo WWTP is located in Sestao (Bizkaia, Spain) with 1.5 million population equivalent. The biological treatment in the water line treats 4 m^3/s and contains six equal lines for organic matter and nitrogen removal with an RDN configuration. Additionally, each of the six lines contains three secondary settlers. All the dimensions and relevant characteristics of the units present in the biological treatment are gathered in Figure 6.7. One of the particularities of the reactors in Galindo WWTP is that the reactors are 9 m tall.

The mathematical model proposed is based on the PWM methodology proposed by Grau *et al.* (2007) and Lizarralde *et al.* (2015) and considers: (1) liquid–gas mass transfer for different components (O_2, CO_2, NH_3, N_2, etc.) that have special relevance when dealing with physicochemical processes and buffer capacity needs to be properly described and (2) different gaseous phases (gas hold-up and off-gas) depending on the aeration strategy in the reactor. A detailed description of the model and parameters used can be found in Lizarralde *et al.* (2018).

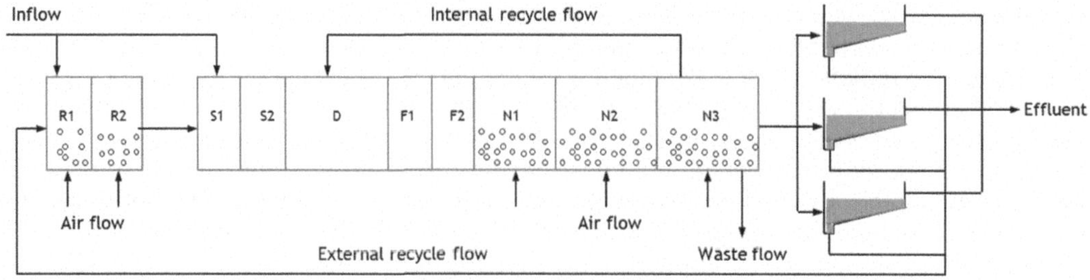

Figure 6.7 Schematic representation of one line of the biological treatment. (*Source*: Lizarralde *et al.*, 2018).

To study the OTE in the WWTP, a model-based study was carried out based on an experimental campaign and the model above introduced. Experimental data were collected from eight experimental campaigns carried out in the aerated reactors in the six lines of the biological treatment. Each of the campaigns measured the efficiency of oxygen transfer in the aerated reactors that depend on three main parameters: difference between clean and dirty water (α), the fouling of the diffusers (F) and the bubble diameter (d_B). The bubble diameter (d_B) was assumed to be 2 mm, which is within the range of expected values for the diffusers present in Galindo WWTP. The parameter αF was adjusted for each reactor to match, as far as possible, the gas flow rate (Q_{ghu}) and the OTE in each of the aerated reactors. Values that minimize the error between experimental and simulated data are presented in Table 6.2. Figure 6.8 shows that the fit between experimental and simulated results is relatively good and αF values obtained with modelling are in accordance with values found in literature (Trillo *et al.*, 2004). From these results, it was concluded that aeration process in Galindo WWTP works properly according to the system installed.

Moreover, Table 6.3 and Figure 6.9 present gas compositions in gas hold-up phases on aerobic and anoxic reactors, which vary depending on the biochemical and chemical processes taking place in the reactors, and calculated *K*la values. The main differences between gas hold-up and off-gas phase (atmosphere) are observed, obviously, in non-aerated reactors where CO_2 and NH_3 composition are considerably higher in comparison with the off-gas due to biological activity taking place in the reactors.

Finally, an exploration by simulation of the effect of the clogging of the diffusers (by means of d_B/F) and submergence of diffusers over OTE is shown. These phenomena affect the transfer efficiency by two mechanisms: on the one hand, the higher the submergence of the diffusers the higher the pressure exerted by the column of water increasing the solubility of oxygen in water. On the other, the higher the column of water above the diffuser, the higher the contact time between the gaseous and aqueous phase. To study the effect of the submergence the reactor height has been changed between 4.5 and 9 m. The results are presented in Figure 6.10.

Table 6.2 αF values in aerated reactors.

	R1	R2	N1	N2	N3
Line 1	0.65	0.50	0.56	0.69	0.62
Line 2	0.44	0.53	0.57	0.59	0.61
Line 3	–	0.61	0.58	0.59	0.58
Line 4	0.43	0.51	0.71	0.69	0.69
Line 5	–	0.53	0.74	0.70	0.67
Line 6	0.58	–	0.64	0.58	0.54

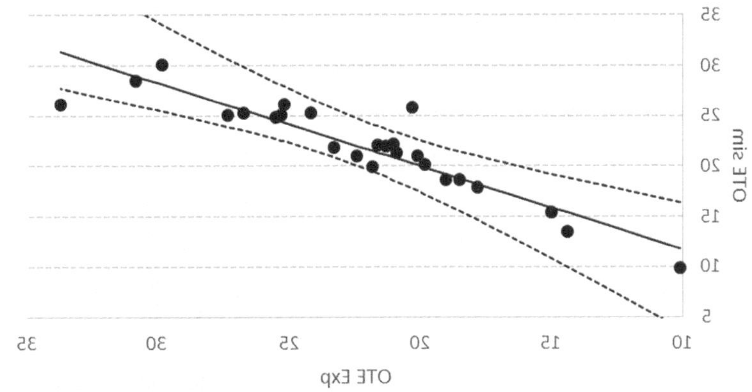

Figure 6.8 Comparison of experimental and simulated OTE.

Table 6.3 $k_L a$ values of the gas hold-up and the off-gas phase of components in all reactors in the Galindo case study.

	Anox		N1		N2		N3	
	Ghu	**Off**	**Ghu**	**Off**	**Ghu**	**Off**	**Ghu**	**Off**
$k_L a_{O2}$	–	24	121	13	102	13	55	9
$k_L a_{CO2}$	0.8	20	106	11	90	11	48	7.8
$k_L a_{N2}$	0.7	20	104	11	88	11	47	7.8

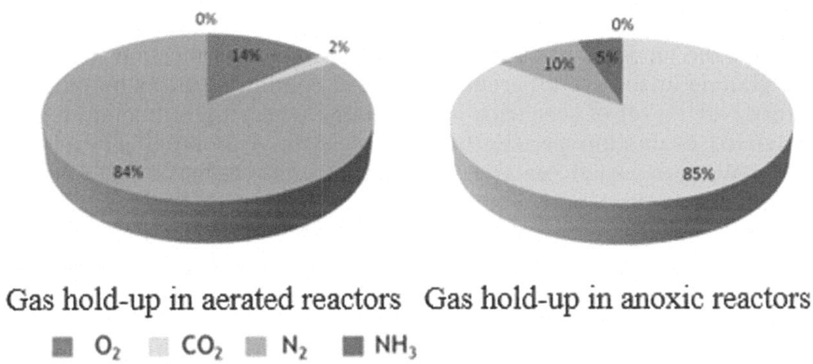

Figure 6.9 Composition of the gas hold-up phase in an aerated reactor and in the anoxic reactor (right).

Figure 6.10 shows that there is an increase of oxygen transfer efficiency the higher the reactor caused by higher contact times between the gaseous and aqueous phases. Also higher reactors exert higher pressures increasing the solubility of the oxygen in the aqueous phase, increasing transfer efficiency. Besides, it is observed that the transfer efficiency is lower when bigger bubbles are formed due to clogging. This effect is strengthened when working in short reactors: when the bubble size increases from 1 to 14 mm the transfer efficiency is 87% lower in short reactors, whereas it is 33%

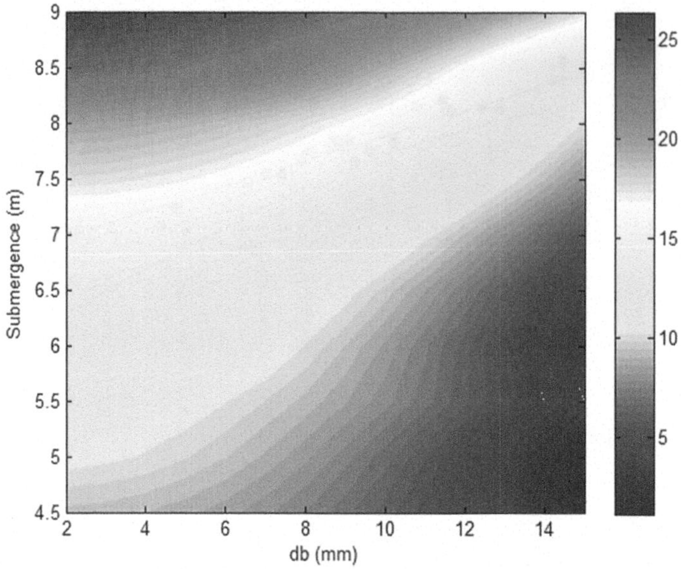

Figure 6.10 Effect of bubble diameter and submergence of diffusers on OTE. (*Source*: Lizarralde *et al.*, 2018).

lower in tall reactors. Thus, the effect of the damage of the state of the diffusers, reflected by an increase in bubble size, and the submergence of the diffusers has been jointly analysed.

6.6 CASE STUDY #5: EFFECT OF INTRA-GRANULE PRECIPITATION IN AN INDUSTRIAL HIGH-RATE INTERNAL CIRCULATION REACTOR

In the last case study the model is used in a non-sludge digester application to assess precipitation within anaerobic granule under two different operational conditions (addition or not of reject water within the reactor). Next, a set of scenarios is formulated where these two strategies are analysed assuming an extended evaluation period (from 3 weeks to 3 months). The plant is located in Kalundborg – North Western part of Zealand (Denmark). Reactor design is based on the BIOPAQ®IC technology (Paques, the Netherlands), a special version of the EGSB concept. In general, EGSB systems employ granular sludge, which is characterized by high mass transfer area, high volumetric conversion rates and ultimately enabling more compact installations. The latter ensures a higher energy recovery potential compared to traditional digestion systems. The reactor is comprised of four parts: (1) a mixing section (M); (2) an expanded sludge bed (R1); (3) a polishing section (R2); and (4) a gas–liquid separator (G–L). In M the influent is evenly distributed within the reactor. The water flows upwards to R1 where most of the organic pollutants are converted into biogas. The effective contact between water and granular anaerobic biomass allows for a higher load. Next, the biogas produced is collected in the lower separation module and flows upwards through the riser to G–L at the top of the reactor. In this section, the water and biogas are separated. The water flows into the downer to the bottom, and it is mixed with the influent. This flow of water from R1 through the riser and the downer and back into M is called the internal circulation (IC) and it gives the reactor its name. In fact, the lifting forces of the collected biogas are used to bring about the IC of the liquid (and granular sludge) over R1. The water from R1 flows to the upper compartment (R2), which has a lower organics concentration. In this section, the rest of the organic material is transformed into biogas. This biogas flows upwards to the headspace, while water is sent to *M* through the downer or

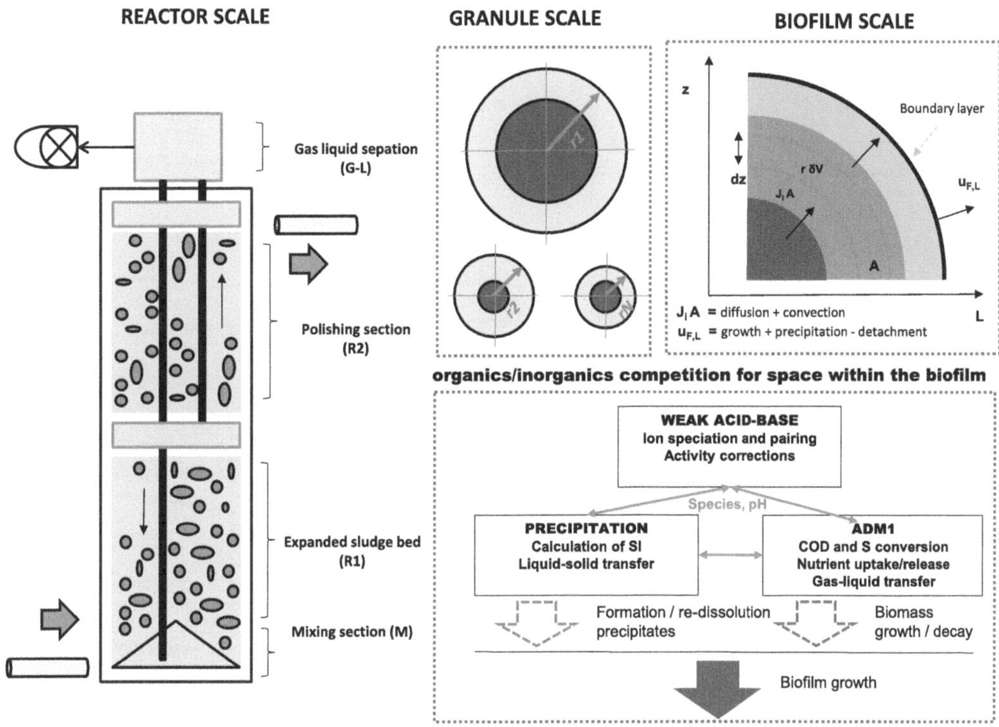

Figure 6.11 Schematic diagram of the reactor of case study #5 and the multiscale modelling approach. (*Source*: Feldman *et al.*, 2019).

leaves the reactor with the overflow. The biogas leaves the reactor at the top and the polished effluent through the effluent pipe (see Figure 6.11 for details).

Multiple mineral precipitation (e.g., $CaCO_3$, $Ca_3(PO_4)_2$) is a common problem in highly loaded anaerobic digestion systems treating industrial wastewater (Batstone *et al.*, 2002b), causing displacement of anaerobic biomass and acting as a barrier to substrate and product transfer. The formation of precipitates at different locations in the reactor (granules, pipes) may cause detrimental (loss of methanogenic activity) or even catastrophic (cementation) effects on reactor performance. While calcium carbonate formation has been assessed in a bulk system using simple extensions to the ADM1 (Batstone & Keller, 2003), it is important to account for in-granule precipitation due to local pH and ion gradients. A multi-scale model, incorporating the acid–base approach of the PCM was developed in Feldman *et al.* (2017). This was subsequently applied to multi-criteria plant-wide optimization in Feldman *et al.* (2018). The work here focuses on prediction and impact of metal ion precipitation incorporating the precipitation model from Chapter 4.

A multi-scale, distributed parameter model-based approach is used to predict multiple mineral precipitation potential (Feldman *et al.*, 2019). At the reactor scale, the mixing section (M), the expanded sludge blanket (R1) and the polishing section (R2) are described as a series of CSTRs. At the granular scale, granule size and distribution is assumed to be dependent on reactor height as a function of TSS measurements. Bigger granules ($L_{max1} = 4$ mm) are located in the bottom (M) of the reactor. The granule radius decreases ($L_{max2} = 2$ mm, $L_{max3} = 0.5$ mm) as a function of reactor height (R1, R2). Finally, at the biofilm level, a one-dimensional model is constructed where the mass balances are derived from spherical coordinates. Reaction rates in both the bulk and the biofilm are evaluated

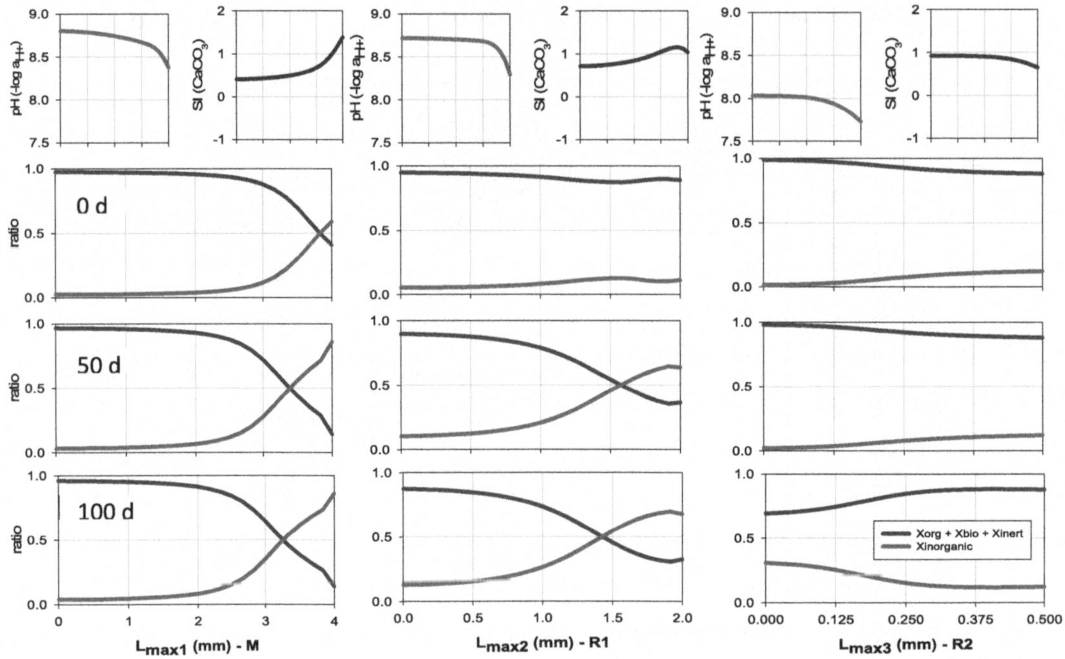

Figure 6.12 Top: pH and SI within the granule at different reactor heights (*M*, *R1*, *R2*). Bottom: Predicted organic ($X_{org} + X_{inert} + X_{bio}$) vs. inorganic ($X_{inorganic}$) contents within the granule at different reactor heights (*M*, *R*1, *R*2: columns 1–3) and simulation times ($t = 0$, $t = 50$, $t = 100$ days: rows 1–3). (*Source*: Feldman *et al.*, 2019).

using the stoichiometry and kinetics as described in the ADM1 extended with P and S (Batstone *et al.*, 2002a; Flores-Alsina *et al.*, 2016). An activity and ion pairing-capable model is required to simulate the behaviour of divalent ions, and importantly, the case presented here is one of the first instances of the use of the PCM in a distributed parameter (1-D) application.

Model predictions suggest that a higher contribution of reject water promotes the potential risk of in-granule $CaCO_3$ formation as a result of the increased quantity of Ca arriving with that stream (lime is used for sludge stabilization) combined with creation of strong pH gradients within the biofilm. The distribution of these precipitates depends on: (a) reactor height; and (b) granule size. In the lower parts of the reactor, precipitates tend to accumulate compared to the upper parts. Also, in larger granules, $CaCO_3$ is expected to form a shell in the outer layers of the granule. On the contrary, in smaller granules precipitates tend to spread all the way to the centre of the granules (see Figure 6.12). Lastly, the study reveals the potential undesirable effects of long-term use of reject water (a decrease in energy recovery by 20%), which are mainly caused by the competition for space between precipitates and biomass, and the reduced buffer capacity (see Figure 6.13).

6.7 OTHER CASES

The framework presented in this book was developed in a number of plant-wide implementations and specific applications, often utilizing the more fundamental (theoretical, lab, or pilot) work referenced in the main report. While we provide the final recommended implementation (and variations), the range of implementations is summarized here for reference, identifying specific applications and modifications as required.

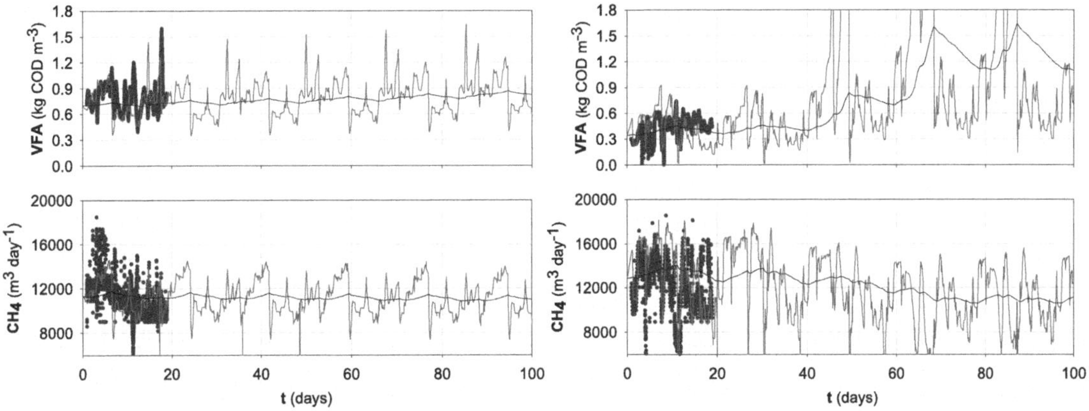

Figure 6.13 Scenario analysis results assuming a long-term (LT) period for both data sets, which corresponds to the different operational modes (addition/no addition of reject water). Markers indicate measurements and red lines are simulated results. A 20-day exponential moving-average filter is included. (*Source*: Feldman *et al.*, 2019).

6.7.1 Urban

While implementation in the hypothetical BSM2 framework is presented in detail in this chapter, this was rapidly implemented in compatible physical full-scale plants. Kazadi Mbamba (Kazadi Mbamba *et al.*, 2016) simulated a large urban treatment plant in South East Queensland, Australia, with a focus on plant-wide simulation of phosphorus chemistry. This required extensive input characterization for cations (as is normally only done for organics, flow and nutrients), and cations and P were validated on a plant-wide basis, with phosphorus recovery scenarios assessed. Aside to this work (using the same plant as a source of inoculum), and focusing on the digester only, the PCM was used to investigate the complex reasons for enhanced phosphorus solubility (and potentially P recovery) at elevated pressures up to 6 bar (Latif *et al.*, 2018). Kazadi-Mbamba further extended the phosphorus analysis to assess iron dosing strategies in a Swedish multi-stage aerobic MBR system (Kazadi Mbamba *et al.*, 2019), identifying that the dosing point has a critical impact on required iron dose to achieve a specific effluent phosphorus level in a closed-loop control system (see case study #2).

Multiple plant layouts were assessed on a competitive basis for the Cape Flats treatment system in South Africa (Flores-Alsina *et al.*, 2021). This incorporated extensive scenario analysis, including atypical climate impacts, centralized biosolids treatment, use of thermal hydrolysis vs. conventional digestion, and was able to fully assess integrated plant effects, including capacity of the mainline system to treat return loads from a centralized thermal hydrolysis digestion system.

6.7.2 Industrial

The same model has been applied to industrial systems. More specifically, Monje *et al.* (2021) tested the prediction capabilities of the model presented in this report when describing the effect of quicklime addition (CaO) on the waste streams of an industrial WTS is assessed. Simulation results show that the model developed within the study is able to reproduce soluble, particulate and total fractions of selected compounds at different plant locations. The main processes included are high pH-induced hydrolysis, precipitation and solid–liquid separation. In Monje *et al.* (2022a,b), the same approach was used to conduct integrated simulations in the largest industrial wastewater treatment plant in Northern Europe. This study also shows that the proposed approach is capable to reproduce main streams neutralization, volatile fatty acid production, particulate removal and nitrate denitrification in the anaerobic water line (buffer tank, primary clarifier, pre-acidification tank). It also correctly predicts organics transformation into methane in the anaerobic granular sludge reactor. Lastly, it is possible to describe biological and chemical N and P removal processes in the activated sludge and the quality of bio-solids after inactivation/dewatering (reject water/cake).

6.7.3 Electrochemical

Electrochemical processes (including microbial electrochemical processes) are common and emerging technologies for water and wastewater treatment. Applications include desalination (Nikonenko *et al.*, 2010), nutrient recovery (Thompson Brewster *et al.*, 2016), and various forms of electrochemical oxidation and reduction, including electrolysis (Shen *et al.*, 2011), as well as applications such as electrocoagulation (where the metal ion is supplied by sacrificial anodes). Electrolysis is emerging as a wastewater reaction to provide oxidants, as well as to collocate hydrogen and oxygen production where the oxygen has value (i.e., a wastewater treatment plant). The area of electrochemical process modelling has been a dispersed field to date, as many models are dedicated to the application, but the underlying processes are the same, and include (a) the electrolyte compartments, commonly separated into a number of compartments by ion-selective membranes, or ion bridges; (b) electrodes, where terminal reactions occur, and (c) the external circuit, which may represent a load or current source depending on the application. In application-specific models, the electrolyte is very well defined (e.g., single salt brine or pure water), and the PCM is parameterized by internal resistance (Shen *et al.*, 2011), or the pH set (e.g., Nikonenko *et al.* (2010)). Spatial and temporal variations will have an enormous impact on key outputs such as whole cell potential, current efficiency, and ion selectivity, and particularly in a wastewater matrix, a more complex electrochemical model is required. This also generally needs to consider multi-dimensional variation, as from inlet to outlet, the system is plug-flow (due to narrow chamber widths), and concentration will vary within electrolyte compartments due to concentration polarization (Thompson Brewster *et al.*, 2016).

These considerations justify the use of an advanced electrochemical model. Ions in an electrochemical cell are subject to ion migration (between electrodes) in addition to diffusion/convection. Migration is proportional to diffusivity of ion, charge of ion, and the potential gradient. However, ion availability is strongly impacted by ion pairing and activity (Thompson Brewster *et al.*, 2016), and variable ion transport (including protons and hydroxide ions) can change both the potential and nature of terminal electrode reactions. Indeed, using the PCM acid–base implementation, including pairing and activity, Thompson Brewster *et al.* (2016) found preferential transport of potassium and ammonium (over calcium and magnesium) during nutrient recovery electrodialysis could be largely related to pairing and activity effects for the divalents, despite having a higher charge than the monovalents. Precipitation was added to assess the relative impact of calcium and struvite scaling (Thompson Brewster *et al.*, 2017b).

6.8 CONCLUSIONS

This chapter presented general applications of how advanced chemistry is applied to promote complex biochemical models. This is not particularly dependent on the underlying biochemical models, which can vary substantially, with the two approaches of PWBM (Grau *et al.*, 2007; Lizarralde *et al.*, 2015), as well as the separate biochemical model (with interface) of the benchmark simulation models (Flores-Alsina *et al.*, 2016; Gernaey *et al.*, 2014) utilizing a very similar physicochemical framework. The chemistry models used are either the reference PCM provided in this book, or very similar, and though complex, have been applied in large systems, with multiple control volumes (including distributed parameter applications). They can also operate in isolation from biochemical models (e.g., for the use in electrochemistry). This indicates that high complexity chemistry models are highly scalable, and produce comparable results across different implementations.

doi: 10.2166/9781780409832_0077

Chapter 7

Conclusions and resources

The generalised approach provided here is intended to operate across the water line. A reference model is provided in the appendix (and available in Github), linked to a number of different applications of mainly ASM (Henze *et al.*, 2000) and ADM1 (Batstone *et al.*, 2002a) biochemical models. The core of the model is the ion speciation and activity model, which is described in Chapter 2. This is largely fixed (by thermodynamics), and common across different implementers (including taskgroup members, but also third-party providers and commercial software). The main variations are in algebraic solver approach. As noted in Chapter 2, this can be applied to any biochemical model, or even simply a chemical component conservation model. A state vector of speciating compounds is fed to the speciation module, and species concentrations, pH, activity, and potentially other relevant properties (e.g., ionic strength) are returned. Some modification of the input state vector may be required for specific applications, as well as inactivation or addition of key components (and related species). This will operate with reasonable accuracy across the water cycle, and key limitations are applicable at very high strengths, including sea-water applications.

Approaches to precipitation and gas transfer vary to a greater degree. While the basic theory remains common, the form of the precipitation equation may consider additional factors (including type of precipitate or colloidal solid). Gas-transfer implementations vary substantially, due to the historical dominance of practical. We have taken a bidirectional and fundamental approach to presentation of gas-transfer theory, with (e.g.) matrix effects being due to changes in gas species activity. This is unlikely (and probably not beneficial) to displace the more empirical, but far more extensively validated alpha–beta–theta approach currently applied for simulating aeration (Stenstrom & Gilbert, 1981). However, where there is not such a well-developed framework (e.g., gas stripping), it is reasonable to take the theoretical approach to mass transfer.

Finally, we note that the science we present here is not novel, but is dispersed amongst both the non-wastewater literature and wastewater engineering. As mentioned in Chapter 1, we note particularly that commercial simulation software has incorporated most of the elements of advanced speciation, precipitation and coagulation, and gas-transfer prior to incorporation of the PCM. As the application of resource recovery advances, particularly processes which are critically dependent on chemistry (adsorption, stripping, membrane-phase separation, electrochemistry), we hope that these fields of research benefit from early incorporation of advanced aquatic chemistry models.

NOMENCLATURE

Symbol	Description	Default Units
γ_i	Activity coefficient component i	ND
$v_{i,j}$	Stoichiometric coefficient for component i in balance or reaction j	Generally integer for aquatic chemistry
Ω	Oversaturation (for precipitates)	ND
θ	Temperature correction factor	
δ	Film thickness	m
ε	Void fraction	ND
μ	Dynamic viscosity	Pa s
ρ	Density	kg m^{-3}
ΔG^0_{rxn}	Free energy of reaction	J $mole^{-1}$
ΔH^0_{rxn}	Enthalpy energy of reaction	J $mole^{-1}$
A	Total area (for mass transfer)	m^2
a	Volume-specific area (for mass transfer)	m^{-1}
$\{A\}$	Activity component A	ND
$\{S_i\}$	Activity component i	ND
$[A]$	Concentration component A	mole L^{-1} (M)
$[S_i]$	Concentration component i	mole L^{-1} (M)
S_i	Dynamic state concentration variable for soluble component i	Varies (commonly mole L^{-1}, mgN L^{-1} or gCOD L^{-1})
Z_i	Implicit algebraic variable for species i	Varies
X_i	Dynamic state concentration variable for particulate component i	Varies
C_i	Dynamic state concentration variable for colloidal component i	Varies
B_0	Particle birth rate	
d_b	Diameter of bubble	m
d_{bs}	Sauter mean diameter	m
D_i	Diffusion coefficient for component i	m^2 s^{-1}
H_i	Henry's law gas solubility for component i	ND or M bar^{-1}
$K_{H,i}$	Henry's law gas volatility ($K_H = 1/H_i$)	ND or bar M^{-1}
I	Ionic strength	Mole L^{-1} (M)
IAP	Ion activity product	
J	Jacobian (multi-dimensional gradient)	N/A
J_i	Flux for component i	g s^{-1} m^{-2}
L	length	m
N_C	Number of components	Integer
N_S	Number of species	Integer
P	Total pressure	bar
p_i	Partial pressure for component i	bar

pH	$-\log\{H^+\}$	ND
$p\varepsilon$	Electron activity ($-\log\{e^-\}$)	ND
pK_a	$-\log(K_a)$	ND
pK_{sp}	$-\log(K_{sp})$	ND
K	Equilibrium constant (varies with temperature and pressure)	Varies (ND when used with activities)
K'	Equilibrium constant corrected for activity	Varies
K''	Equilibrium constant corrected for activity with other simplifications (see Equation (2.31))	Varies (ND when used with activities)
K_a	Acid–base equilibrium constant	Varies (ND when used with activities)
K_{sp}	Solubility product equilibrium constant	
k	Kinetic coefficient	Varies
k_L	Liquid side gas-transfer coefficient	$m\ s^{-1}$
k_G	Gas side gas-transfer coefficient	$m\ s^{-1}$ or $m\ M\ s^{-1}\ bar^{-1}$
R	Gas law constant	Depends on context
r_i	Reaction rate for component or process i	
Sc	Schmidt number	ND
T	Temperature	Kelvin or °C
t	Time	s or d
V_G	Volume of gas	m^3
V_L	Volume of liquid	m^3
v_r	Bubble slip velocity	$m\ s^{-1}$
z_i	Valence for component i	Integer

SOFTWARE RESOURCES AND REFERENCE MODEL

The reference model is provided in conjunction with various implementations of IWA biochemical models (mainly ASM series and ADM): https://github.com/wwtmodels/Plant-Wide-Models

doi: 10.2166/9781780409832_0081

References

Abbona F. and Boistelle R. (1979). Growth morphology and crystal habit of struvite crystals ($MgNH_4PO_4$ $6H_2O$). *Journal of Crystal Growth*, **46**(3), 339–354.

Allison J. D., Brown D. S., Novo-Gradac K. J. A. and GA, 30605: EPA (USA). (1991). MINTEQA2/PRODEF2, A Geochemical Assessment Model for Environmental Systems: Version 3.0 User's Manual, EPA/600/3-91/021, Athens.

Amaral A., Gillot S., Garrido-Baserba M., Filali A., Karpinska A. M., Plósz B. G., De Groot C., Bellandi G., Nopens I., Takács I., Lizarralde I., Jimenez J. A., Fiat J., Rieger L., Arnell M., Andersen M., Jeppsson U., Rehman U., Fayolle Y., Amerlinck Y. and Rosso D. (2019). Modelling gas–liquid mass transfer in wastewater treatment: when current knowledge needs to encounter engineering practice and vice versa. *Water Science and Technology*, **80**(4), 607–619.

Amerlinck Y., De Keyser W., Urchegui G. and Nopens I. (2016). A realistic dynamic blower energy consumption model for wastewater applications. *Water Science and Technology*, **74**(7), 1561–1576.

Appelo C. A. J. and Postma D. (2005). Geochemistry, Groundwater and Pollution. CRC Press.

Ariyanto E., Sen T. K. and Ang H. M. (2014). The influence of various physico-chemical process parameters on kinetics and growth mechanism of struvite crystallisation. *Advanced Powder Technology*, **25**(2), 682–694.

ASCE. (2007). Measurement of Oxygen Transfer in Clean Water.

Baeten J. E., van Loosdrecht M. C. M. and Volcke E. I. P. (2020). When and why do gradients of the gas phase composition and pressure affect liquid-gas transfer? *Water Research*, **178**, 115844.

Banerjee A. (2016). Estimation of dolomite formation: Dolomite precipitation and dolomitization. *Journal of the Geological Society of India*, **87**(5), 561–572.

Barat R., Montoya T., Seco A. and Ferrer J. (2011). Modelling biological and chemically induced precipitation of calcium phosphate in enhanced biological phosphorus removal systems. *Water Research*, **45**(12), 3744–3752.

Batstone D. J. and Keller J. (2003). Industrial applications of the IWA anaerobic digestion model No. 1 (ADM1). *Water Science and Technology*, **47**(12), 199–206.

Batstone D. J., Keller J., Angelidaki I., Kalyuzhnyi S. V., Pavlostathis S. G., Rozzi A., Sanders W. T. M., Siegrist H. and Vavilin V. A. (2002a). Anaerobic Digestion Model No. 1 (ADM1), IWA Task Group for Mathematical Modelling of Anaerobic Digestion Processes. IWA Publishing, London.

Batstone D. J., Landelli J., Saunders A., Webb R. I., Blackall L. L. and Keller J. (2002b). The influence of calcium on granular sludge in a full-scale UASB treating paper mill wastewater. *Water Science and Technology*, **45**(10), 187–193.

Batstone D. J., Amerlinck Y., Ekama G., Goel R., Grau P., Johnson B., Kaya I., Steyer J. P., Tait S., Takaćs I., Vanrolleghem P. A., Brouckaert C. J. and Volcke E. (2012). Towards a generalized physicochemical framework. *Water Science and Technology*, **66**(6), 1147–1161.

Batstone D. J., Puyol D., Flores-Alsina X. and Rodríguez J. (2015). Mathematical modelling of anaerobic digestion processes: applications and future needs. *Reviews in Environmental Science and Biotechnology*, **14**(4), 595–613.

Batstone D. J., Flores-Alsina X. and Hauduc H. (2018). Modeling the phosphorus cycle in the wastewater treatment process. In: C. Schaum (ed.) Phosphorus: Polluter and Resource of the Future – Removal and Recovery from Wastewater. IWA Publishing, London, 219–234.

Bencsik D., Takács I. and Rosso D. (2022). Dynamic alpha factors: Prediction in time and evolution along reactors. *Water Research*, **216**, 118339.

Bhuiyan M. I., Mavinic D. S. and Koch F. A. (2008). Phosphorus recovery from wastewater through struvite formation in fluidized bed reactors: a sustainable approach. *Water Science and Technology*, **57**(2), 175–181.

Billet R. and Schultes M. (1993). Predicting mass transfer in packed columns. *Chemical Engineering & Technology*, **16**(1), 1–9.

Boistelle R., Abbona F. and Lundager Madsen H. E. (1983). On the transformation of struvite into newberyite in aqueous systems. *Physics and Chemistry of Minerals*, **9**(5), 216–222.

Bouropoulos N. C. and Koutsoukos P. G. (2000). Spontaneous precipitation of struvite from aqueous solutions. *Journal of Crystal Growth*, **213**(3), 381–388.

Braunstein P. (2010). Hollemann, Wiberg & Wiberg: Lehrbuch der Anorganischen Chemie. Von Hollemann-Wiberg. Walter de Gruyter Verlag & Co., Berlin – New York, 1984. XXX, 1451 S., 290 Abb., 112 Tab., DM 120,–. ISBN 3-11-007511-3. Nachrichten aus Chemie, Technik und Laboratorium 33.

Brečević L. and Kralj D. (2007). On Calcium Carbonates: from Fundamental Research to Application, p. 467, Hrvatsko kemijsko društvo.

Brečević L. and Nielsen A. E. (1989). Solubility of amorphous calcium carbonate. *Journal of Crystal Growth*, **98**(3), 504–510.

Brouckaert C. J., Brouckaert B. M. and Ekama G. A. (2021). Integration of complete elemental mass-balanced stoichiometry and aqueous-phase chemistry for bioprocess modelling of liquid and solid waste treatment systems – Part 1: the physico-chemical framework. *Water SA*, **47**, 276–288.

Brouckaert C. J., Brouckaert B. M. and Ekama G. A. (2022). Integration of complete elemental mass-balanced stoichiometry and aqueous-phase chemistry for bioprocess modelling of liquid and solid waste treatment systems – Part 5: Ionic speciation. *Water SA*, **48**, 32–39.

Calderbank P. H. and Moo-Young M. B. (1961). The continuous phase heat and mass-transfer properties of dispersions. *Chemical Engineering Science*, **16**(1), 39–54.

Cohen E. R., Cvitas T., Frey J., Holmstrom B., Kuchitsu K., Marquardt R., Mills I., Pavese F., Quack M., Stohner J., Strauss H., Takami M. and Thor A. J. (2007). Quantities, Units and Symbols in Physical Chemistry. IUPAC Green Book, Cambridge, UK.

Cussler E. L. (2009). Diffusion: Mass Transfer in Fluid Systems. Cambridge University Press, Cambridge.

Daelman M. R. J., Van Eynde T., van Loosdrecht M. C. M. and Volcke E. I. P. (2014). Effect of process design and operating parameters on aerobic methane oxidation in municipal WWTPs. *Water Research*, **66**, 308–319.

Danckwerts P. V. (1951). Significance of liquid-film coefficients in gas absorption. *Industrial & Engineering Chemistry*, **43**(6), 1460–1467.

Doyle J. D. and Parsons S. A. (2002). Struvite formation, control and recovery. *Water Research*, **36**(16), 3925–3940.

Durrant A. E., Scrimshaw M. D., Stratful I. and Lester J. N. (1999). Review of the feasibility of recovering phosphate from wastewater for use as a raw material by the phosphate industry. *Environmental Technology*, **20**(7), 749–758.

Elduayen-Echave B., Azcona M., Grau P. and Schneider P. A. (2020). Effect of the shear rate and supersaturation on the nucleation and growth of struvite in batch stirred tank reactors. *Journal of Water Process Engineering*, **38**, 101657.

Elduayen-Echave B., Lizarralde I., Schneider P., Ayesa E., Larraona G. and Grau P. (2021). Inclusion of shear rate effects in the kinetics of a discretized population balance model: Application to struvite precipitation. *Water Research*, **200**, 117242.

Fairlamb M., Jones R., Takács I. and Bye C. (2003). Formulation of a general model for simulation of pH in wastewater treatment processes. Proceedings of WEFTEC, 2003, Los Angeles, California.

Feldman H., Flores-Alsina X., Kjellberg K., Jeppsson U., Batstone D. J. and Gernaey K. V. (2018). Model-based analysis and optimization of a full-scale industrial high-rate anaerobic bioreactor. *Biotechnology and Bioengineering*, **115**(11), 2726–2739.

Feldman H., Flores-Alsina X., Ramin P., Kjellberg K., Jeppsson U., Batstone D. J. and Gernaey K. V. (2017). Modelling an industrial anaerobic granular reactor using a multi-scale approach. *Water Research*, **126**, 488–500.

Feldman H., Flores-Alsina X., Ramin P., Kjellberg K., Jeppsson U., Batstone D. J. and Gernaey K. V. (2019). Assessing the effects of intra-granule precipitation in a full-scale industrial anaerobic digester. *Water Science and Technology*, **79**(7), 1327–1337.

Ferguson J. F. and King T. (1977). A model for aluminum phosphate precipitation. *Journal (Water Pollution Control Federation)*, **49**(4), 646–658.

Fernández-Arévalo T., Lizarralde I., Fdz-Polanco F., Pérez-Elvira S. I., Garrido J. M., Puig S., Poch M., Grau P. and Ayesa E. (2017). Quantitative assessment of energy and resource recovery in wastewater treatment plants based on plant-wide simulations. *Water Research*, **118**, 272–288.

Fernández-Arévalo T., Lizarralde I., Grau P. and Ayesa E. (2014). New systematic methodology for incorporating dynamic heat transfer modelling in multi-phase biochemical reactors. *Water Research*, **60**, 141–155.

Fiat J., Filali A., Fayolle Y., Bernier J., Rocher V., Spérandio M. and Gillot S. (2019). Considering the plug-flow behavior of the gas phase in nitrifying BAF models significantly improves the prediction of N_2O emissions. *Water Research*, **156**, 337–346.

Flores-Alsina X., Kazadi Mbamba C., Solon K., Vrecko D., Tait S., Batstone D. J., Jeppsson U. and Gernaey K. V. (2015). A plant-wide aqueous phase chemistry module describing pH variations and ion speciation/pairing in wastewater treatment process models. *Water Research*, **85**, 255–265.

Flores-Alsina X., Ramin E., Ikumi D., Harding T., Batstone D., Brouckaert C., Sotemann S. and Gernaey K. V. (2021). Assessment of sludge management strategies in wastewater treatment systems using a plant-wide approach. *Water Research*, **190**, 116714

Flores-Alsina X., Solon K., Kazadi Mbamba C., Tait S., Gernaey K. V., Jeppsson U. and Batstone D. J. (2016). Modelling phosphorus (P), sulfur (S) and iron (Fe) interactions for dynamic simulations of anaerobic digestion processes. *Water Research*, **95**, 370–382.

Frossling N. (1938). Uber die Verdunstung fallender Tropfen.

Galbraith S. C. and Schneider P. A. (2014). Modelling and simulation of inorganic precipitation with nucleation, crystal growth and aggregation: a new approach to an old method. *Chemical Engineering Journal*, **240**, 124–132.

Generalic E. (2020). Solubility product constants. EniG. Periodic Table of the Elements. KTF-Split.

Gernaey K. V., Jeppsson U., Vanrolleghem P. A. and Copp J. B. (2014). Benchmarking of Control Strategies for Wastewater Treatment Plants. IWA Scientific and Technical Report No. 23. IWA Publishing, London, UK.

Gillot S. and Héduit A. (2008). Prediction of alpha factor values for fine pore aeration systems. *Water Science and Technology*, **57**(8), 1265–1269.

Gillot S., Capela-Marsal S., Roustan M. and Héduit A. (2005). Predicting oxygen transfer of fine bubble diffused aeration systems – model issued from dimensional analysis. *Water Research*, **39**(7), 1379–1387.

Gough B. and Priedhorsky R. (2017). Multidimensional Root-Finding. In: M. Galassi and J. Theiler (eds). GNU scientific library (GSL) documentation. https://www.gnu.org/software/gsl/doc/html/multiroots.html (accessed 8/2022)

Grau P., de Gracia M., Vanrolleghem P. A. and Ayesa E. (2007). A new plant-wide modelling methodology for WWTPs. *Water Research*, **41**(19), 4357–4372.

Guggenheim E. A. and Turgeon J. C. (1955). Specific interaction of ions. *Transactions of the Faraday Society*, **51**(0), 747–761.

Gutierrez O., Park D., Sharma K. R. and Yuan Z. (2010). Iron salts dosage for sulfide control in sewers induces chemical phosphorus removal during wastewater treatment. *Water Research*, **44**(11), 3467–3475.

Hamann C. H., Hamnett A. and Vielstich W. (2007). Electrochemistry. Wiley-VCH, New York.

Hangos K. and Cameron I. T. (2001). Process Modelling and Model Analysis. Academic Press, San Diego.

Hao T.-W., Xiang P.-Y., Mackey H. R., Chi K., Lu H., Chui H.-K., van Loosdrecht M. C. M. and Chen G.-H. (2014). A review of biological sulfate conversions in wastewater treatment. *Water Research*, **65**, 1–21.

Harned H. S. and Owen B. B. (1958). The Physical Chemistry of Electrolytic Solutions/by Herbert S. Harned and Benton B. Owen, New York: Reinhold Pub. Corp., New York.

Hauduc H., Al-Omari A., Wett B., Jimenez J., De Clippeleir H., Rahman A., Wadhawan T. and Takacs I. (2018). Colloids, flocculation and carbon capture – a comprehensive plant-wide model. *Water Science and Technology*, **79**(1), 15–25.

Hauduc H., Takács I., Smith S., Szabo A., Murthy S., Daigger G. T. and Spérandio M. (2015). A dynamic physicochemical model for chemical phosphorus removal. *Water Research*, **73**, 157–170.

Hauduc H., Wadhawan T., Johnson B., Bott C., Ward M. and Takács I. (2019). Incorporating sulfur reactions and interactions with iron and phosphorus into a general plant-wide model. *Water Science and Technology*, **79**(1), 26–34.

Hem J. D. (1961). Stability field diagrams as aids in iron chemistry studies. *Journal AWWA*, **53**(2), 211–232.

Henze M., Gujer W., Mino T. and van Loosdrecht M. (2000). Activated Sludge Models ASM1, ASM2, ASM2d, and ASM3, IWA Scientific and Technical Report, London.

Higbie R. (1935). The rate of absorption of a pure gas into a still liquid during a short time of exposure. *Transactions of the American Institute of Chemical Engineering*, **31**, 365–389.

Huber P., Neyret C. and Fourest E. (2017). Implementation of the anaerobic digestion model (ADM1) in the PHREEQC chemistry engine. *Water Science and Technology*, **76**(5–6), 1090–1103.

Jeppsson U., Alex J., Batstone D. J., Benedetti L., Comas J., Copp J. B., Corominas L., Flores-Alsina X., Gernaey K. V., Nopens I., Pons M. N., Rodríguez-Roda I., Rosen C., Steyer J. P., Vanrolleghem P. A., Volcke E. I. P. and Vrecko D. (2013). Benchmark simulation models, quo vadis? *Water Science and Technology*, **68**(1), 1–15.

Jiang L.-M., Garrido-Baserba M., Nolasco D., Al-Omari A., DeClippeleir H., Murthy S. and Rosso D. (2017). Modelling oxygen transfer using dynamic alpha factors. *Water Research*, **124**, 139–148.

Jones A. G. (2002). Crystallization Process Systems. Butterworth-Heinemann, Oxford, UK. ISBN 9780750655200.

Kazadi Mbamba C. (2016). Dynamic modelling of minerals precipitation in wastewater treatment.

Kazadi Mbamba C., Batstone D. J., Flores-Alsina X. and Tait S. (2015a). A generalised chemical precipitation modelling approach in wastewater treatment applied to calcite. *Water Research*, **68**, 342–353.

Kazadi Mbamba C., Tait S., Flores-Alsina X. and Batstone D. J. (2015b). A systematic study of multiple minerals precipitation modelling in wastewater treatment. *Water Research*, **85**, 359–370.

Kazadi Mbamba C., Flores-Alsina X., John Batstone D. and Tait S. (2016). Validation of a plant-wide phosphorus modelling approach with minerals precipitation in a full-scale WWTP. *Water Research*, **100**, 169–183.

Kazadi Mbamba C., Lindblom E., Flores-Alsina X., Tait S., Anderson S., Saagi R., Batstone D. J., Gernaey K. V. and Jeppsson U. (2019). Plant-wide model-based analysis of iron dosage strategies for chemical phosphorus removal in wastewater treatment systems. *Water Research*, **155**, 12–25.

Khawaga R. I., Al-Asheh S., Jabbar N. A. and Abouleish M. (2018). Modeling and validation of chlorination breakpoint with nitrite in wastewater treatment. *Water Quality Research Journal*, **53**(4), 219–230.

Kip N. and van Veen J. A. (2015). The dual role of microbes in corrosion. *The ISME Journal*, **9**(3), 542–551.

Kralj D., Brečević L. and Kontrec J. (1997). Vaterite growth and dissolution in aqueous solution III. Kinetics of transformation. *Journal of Crystal Growth*, **177**(3–4), 248–257.

Langmuir D. (1997). Aqueous Environmental Geochemistry. Prentice-Hall, Upper Saddle River, New Jersey.

Latif M. A., Mehta C. M. and Batstone D. J. (2018). Enhancing soluble phosphate concentration in sludge liquor by pressurised anaerobic digestion. *Water Research*, **145**, 660–666.

Le Corre K. S., Valsami-Jones E., Hobbs P. and Parsons S. A. (2009). Phosphorus recovery from wastewater by struvite crystallization: a review. *Critical Reviews in Environmental Science and Technology*, **39**(6), 433–477.

Lewis W. K. and Whitman W. G. (1924). Principles of gas absorption. *Industrial & Engineering Chemistry*, **16**(12), 1215–1220.

Li B. and Bishop P. L. (2001). The application of ORP in activated sludge wastewater treatment processes. *Environmental Engineering Science*, **18**(5), 309–321.

Lindberg R. D. and Runnells D. D. (1984). Ground water redox reactions: an analysis of equilibrium state applied to Eh measurements and geochemical modeling. *Science*, **225**(4665), 925–927.

Lizarralde I., Fernández-Arévalo T., Brouckaert C., Vanrolleghem P., Ikumi D. S., Ekama G. A., Ayesa E. and Grau P. (2015). A new general methodology for incorporating physico-chemical transformations into multi-phase wastewater treatment process models. *Water Research*, **74**, 239–256.

Lizarralde I., Fernández-Arévalo T., Beltrán S., Ayesa E. and Grau P. (2018). Validation of a multi-phase plant-wide model for the description of the aeration process in a WWTP. *Water Research*, **129**, 305–318.

Lizarralde I., Fernández-Arévalo T., Manas A., Ayesa E. and Grau P. (2019). Model-based optimization of phosphorus management strategies in Sur WWTP, Madrid. *Water Research*, **153**, 39–52.

Loewenthal R. E., Kornmüller U. R. C. and van Heerden E. P. (1994). Modelling struvite precipitation in anaerobic treatment systems. *Water Science and Technology*, **30**(12), 107.

López L. R., Dorado A. D., Mora M., Gamisans X., Lafuente J. and Gabriel D. (2016). Modeling an aerobic biotrickling filter for biogas desulfurization through a multi-step oxidation mechanism. *Chemical Engineering Journal*, **294**, 447–457.

López L. R., Mora M., Van der Heyden C., Baeza J. A., Volcke E. and Gabriel D. (2021). Model-based analysis of feedback control strategies in aerobic biotrickling filters for biogas desulfurization. *Processes*, **9**(2), 208.

Maharaj B. C., Mattei M. R., Frunzo L., Hullebusch E. D. V. and Esposito G. (2019). ADM1 based mathematical model of trace element complexation in anaerobic digestion processes. *Bioresource Technology*, **276**, 253–259.

Massman W. J. (1998). A review of the molecular diffusivities of H_2O, CO_2, CH_4, CO, O_3, SO_2, NH_3, N_2O, NO, and NO_2 in air, O_2 and N_2 near STP. *Atmospheric Environment*, **32**(6), 1111–1127.

Matthews M. A. (2000). A to Z of Thermodynamics by Pierre Perrot (Université des sciences et technologies de Lille). Oxford University Press, Oxford, New York, and Tokyo. 1998. vi + 329 pp. $65.00. ISBN 0-19-856556-9 (Hardback). Journal of the American Chemical Society 122(15), 3799-3800.

Merkel B. J., Nordstrom D. K. and Planer-Friedrich B. (2008). Groundwater Geochemistry: A Practical Guide to Modeling of Natural and Contaminated Aquatic Systems. Springer, Berlin Heidelberg.

Meyer J. L. and Weatherall C. C. (1982). Amorphous to crystalline calcium phosphate phase transformation at elevated pH. *Journal of Colloid and Interface Science*, **89**(1), 257–267.

Mohd Pu'ad N. A. S., Abdul Haq R. H., Mohd Noh H., Abdullah H. Z., Idris M. I. and Lee T. C. (2020). Synthesis method of hydroxyapatite: a review. *Materials Today: Proceedings*, **29**, 233–239.

Monje V., Nobel P., Junicke H., Kjellberg K., Gernaey K. V. and Flores-Alsina X. (2021). Assessment of alkaline stabilization processes in industrial waste streams using a model-based approach. *Journal of Environmental Management*, **293**, 112806.

Monje V., Junicke H., Batstone D. J., Kjellberg K., Gernaey K. V. and Flores-Alsina X. (2022a). Predicting mass and volumetric flow in a full-scale waste treatment plant. *Chemical Engineering Journal*, **445**, 136774.

Monje V., Owsianiak M., Junicke H., Kjellberg K., Gernaey K. V. and Flores-Alsina X. (2022b). Economic, technical, and environmental evaluation of retrofitting scenarios in a full-scale industrial wastewater treatment system. *Water Research*, **223**, 118997.

Monteiro L., Figueiredo D., Dias S., Freitas R., Covas D., Menaia J. and Coelho S. T. (2014). Modeling of chlorine decay in drinking water supply systems using EPANET MSX. *Procedia Engineering*, **70**, 1192–1200.

Morel F. M. M. and Hering J. G. (1993). Principles and Applications of Aquatic Chemistry. Wiley-Interscience, New York.

Moussa D. T., El-Naas M. H., Nasser M. and Al-Marri M. J. (2017). A comprehensive review of electrocoagulation for water treatment: Potentials and challenges. *Journal of Environmental Management*, **186**, 24–41.

Mullin J. W. (1992). Crystallization. Butterworth-Heinemann Publications, Ipswich, UK.

Mullin J. W. (2001). Crystallization. Elsevier Butterworth-Heinemann, Oxford.

Munz C. and Roberts P. V. (1984). The Ratio of Gas-Phase to Liquid-Phase Mass Transfer Coefficients in Gas-Liquid Contacting Processes. In: W. Brutsaert and G. H. Jirka (eds), Gas transfer at water surfaces. Springer Netherlands: Dordrecht, pp. 35–45.

Musvoto E., Wentzel M., Loewenthal R. and Ekama G. (2000a). Integrated chemical-physical processes modelling – I. Development of a kinetic-based model for mixed weak acid/base systems. *Water Research*, **34**(6), 1857–1867.

Musvoto E., Wentzel M., Loewenthal R. and Ekama G. (2000b). Integrated chemical-physical processes modelling – II. Simulating aeration treatment of anaerobic digester supernatants. *Water Research*, **34**(6), 1868–1880.

Myerson A. (1993). Handbook of Crystallization. Elsevier Science & Technology Butterworth-Heinemann, Boston, USA.

NFEN. (2004). European Standard: wastewater treatment plants – Part 15: measurement of the oxygen transfer in clean water in activated sludge aeration tanks.

Nielsen A. (1970). Precipitation. *Croatica Chemica Acta*, **42**(2), 319–333.

Nikonenko V. V., Pismenskaya N. D., Belova E. I., Sistat P., Huguet P., Pourcelly G. and Larchet C. (2010). Intensive current transfer in membrane systems: modelling, mechanisms and application in electrodialysis. *Advances in Colloid and Interface Science*, **160**(1), 101–123.

O'Connor D. J. and Dobbins W. E. (1956). The mechanism of reaeration in natural streams. *Journal of the Sanitary Engineering Division*, **82**(6), 1–30.

Ohlinger K. N. P. E., Young T. M. and Schroeder E. D. (1999). Kinetics effects on preferential struvite accumulation in wastewater. *Journal of Environmental Engineering (New York, N.Y.)*, **125**(8), 730–737.

Onda K., Takeuchi H. and Okumoto Y. (1968). Mass transfer coefficients between gas and liquid phases in packed columns. *Journal of Chemical Engineering of Japan*, **1**(1), 56–62.

Pan H., Liu X. Y., Tang R. and Xu H. Y. (2010). Mystery of the transformation from amorphous calcium phosphate to hydroxyapatite. *Chemical Communications*, **46**(39), 7415–7417.

Parkhurst D. L. and Appelo C. A. J. (1999). User's Guide to PHREEQC (Version 2) – a Computer Program for Speciation, Batch Reaction, One-dimensional Transport, and Inverse Geochemical Calculations: U.S. Geological Survey Water – Resources Investigations.

Parkhurst D. L. and Appelo C. A. J. (2013). U.S. Geological Survey Techniques and Methods, book 6, chap. A43 available only at https://pubs.usgs.gov/tm/06/a43/., p. 497 p., U.S. Geological Survey Water – Resources Investigations.

Pitzer K. (1991). Activity Coefficients in Electrolyte Solutions, 2nd edn. CRC Press, Boca Raton.

Plummer L. N. and Busenberg E. (1982). The solubilities of calcite, aragonite and vaterite in CO_2-H_2O solutions between 0 and 90°C, and an evaluation of the aqueous model for the system $CaCO_3$-CO_2-H_2O. *Geochimica et Cosmochimica Acta*, **46**(6), 1011–1040.

Prata A. A., Santos J. M., Timchenko V. and Stuetz R. M. (2018). A critical review on liquid-gas mass transfer models for estimating gaseous emissions from passive liquid surfaces in wastewater treatment plants. *Water Research*, **130**, 388–406.

Puigdomenech I., Plyasunov A. V., Rard J. A. and Grenthe I. (1997). Temperature corrections to thermodynamic data and enthalpy calculations. In: I. Grenthe and I. Puigdomenech (eds). Modelling in Aquatic Chemistry. OECD Nuclear Energy Agency, (http://www.nea.fr): Paris, pp. 427–493.

Recht H. L. and Ghassemi M. (1970). Kinetics and Mechanism of Precipitation and Nature of Precipitate Obtained in Phosphate Removal from Wastewater Using Aluminum and Iron Salts. Report No. 17010 EK1, Federal Water Quality Administration.

Reeves C. R. (1995). Modern Heuristic Techniques for Combinatorial Problems. McGraw-Hill, New York.

Reichert P., Borchardt D., Henze M., Rauch S., Shanahan P., Somlyody L. and Vanrolleghem P. A. (2001). River Water Quality Model No.1: Scientific & Technical Report No. 12. IWA Publishing, London.

Rosen C., Vrecko D., Gernaey K. V., Pons M. N. and Jeppsson U. (2006). Implementing ADM1 for plant-wide benchmark simulations in Matlab/Simulink. *Water Science and Technology*, **54**(4), 11–19.

Roustan M. (2003). Transferts gaz-liquide dans les procédés de traitement des eaux et des effluents gazeux. Lavoisier Tec & Doc, Paris.

Runtti H., Luukkonen T., Niskanen M., Tuomikoski S., Kangas T., Tynjälä P., Tolonen E.-T., Sarkkinen M., Kemppainen K., Rämö J. and Lassi U. (2016). Sulphate removal over barium-modified blast-furnace-slag geopolymer. *Journal of Hazardous Materials*, **317**, 373–384.

Sander R. (2015). Compilation of Henry's law constants (version 4.0) for water as solvent. *Atmospheric Chemistry and Physics*, **15**(8), 4399–4981.

Sawyer C. N., McCarty P. L. and Parkin G. F. (2003). Chemistry for Environmental Engineering and Science. McGraw-Hill, Boston.

Serralta J., Ferrer J., Borrás L. and Seco A. (2004). An extension of ASM2d including pH calculation. *Water Research*, **38**(19), 4029–4038.

Shaddel S., Ucar S., Andreassen J.-P. and Østerhus S. W. (2019). Enhancing efficiency and economics of phosphorus recovery process by customizing the product based on sidestream characteristics – an alternative phosphorus recovery strategy. *Water Science and Technology*, **79**(9), 1777–1789.

Sharma M. M. and Mashelkar R. A. (1968). Absorption with reaction in bubble columns. *Institution of Chemical Engineers Symposium Series*, **28**, 10–21.

Shen M., Bennett N., Ding Y. and Scott K. (2011). A concise model for evaluating water electrolysis. *International Journal of Hydrogen Energy*, **36**(22), 14335–14341.

Shulman H. L., Ullrich C. F., Proulx A. Z. and Zimmerman J. O. (1955). Performance of packed columns. II. Wetted and effective-interfacial areas, gas- and liquid-phase mass transfer rates. *AIChE Journal*, **1**(2), 253–258.

Smith S., Takács I., Murthy S., Daigger G. T. and Szabó A. (2008). Phosphate complexation model and its implications for chemical phosphorus removal. *Water Environment Research*, **80**(5), 428–438.

Solon K., Flores-Alsina X., Mbamba C. K., Volcke E. I. P., Tait S., Batstone D., Gernaey K. V. and Jeppsson U. (2015). Effects of ionic strength and ion pairing on (plant-wide) modelling of anaerobic digestion. *Water Research*, **70**, 235–245.

Solon K., Flores-Alsina X., Kazadi Mbamba C., Ikumi D., Volcke E. I. P., Vaneeckhaute C., Ekama G., Vanrolleghem P. A., Batstone D. J., Gernaey K. V. and Jeppsson U. (2017). Plant-wide modelling of phosphorus transformations in wastewater treatment systems: impacts of control and operational strategies. *Water Research*, **113**, 97–110.

Sötemann S. W., Musvoto E. V., Wentzel M. C. and Ekama G. A. (2005). Integrated biological, chemical and physical processes kinetic modelling Part 1 – Anoxic-aerobic C and N removal in the activated sludge system. *Water SA*, **31**(4), 529–544.

Stenstrom M. K. and Gilbert R. G. (1981). Effects of alpha, beta and theta factor upon the design, specification and operation of aeration systems. *Water Research*, **15**(6), 643–654.

Stumm W. and Morgan J. J. (1996). Aquatic Chemistry: Chemical Equilibria and Rates in Natural Waters. John Wiley and Sons, New York.

Szabó A., Takács I., Murthy S., Daigger G., Licskó I. and Smith S. (2008). Significance of design and operational variables in chemical phosphorus removal. *Water Environment Research: a Research Publication of the Water Environment Federation*, **80**, 407–416.

Tait S., Clarke W. P., Keller J. and Batstone D. J. (2009). Removal of sulfate from high-strength wastewater by crystallisation. *Water Research*, **43**(3), 762–772.

Tait S., Solon K., Volcke E. I. P. and Batstone D. J. (2012). A unified approach to modelling wastewater chemistry, 3rd IWA/WEF Wastewater treatment modelling seminar, 26–28 Feb 2012, In: P. Vanrolleghem (ed.), IWA, Mont-Sainte-Anne, Quebec, Canada.

Takacs I. (2008). Experiments in Activated Sludge Modelling. University of Ghent, Ghent.

Tchobanoglous G., Burton F. and Stensel H. (2003). Metcalf and Eddy Inc. Wastewater Engineering, Treatment and Reuse. McGraw-Hill, New York, NY (US).

Thompson Brewster E., Mehta C. M., Radjenovic J. and Batstone D. J. (2016). A mechanistic model for electrochemical nutrient recovery systems. *Water Research*, **94**, 176–186.

Thompson Brewster E., Jermakka J., Freguia S. and Batstone D. J. (2017a). Modelling recovery of ammonium from urine by electro-concentration in a 3-chamber cell. *Water Research*, **124**, 210–218.

Thompson Brewster E., Ward A. J., Mehta C. M., Radjenovic J. and Batstone D. J. (2017b). Predicting scale formation during electrodialytic nutrient recovery. *Water Research*, **110**, 202–210.

Tobiason J. E., Bazilio A., Goodwill J., Mai X. and Nguyen C. (2016). Manganese removal from drinking water sources. *Current Pollution Reports*, **2**(3), 168–177.

Torres M. A., West A. J. and Li G. (2014). Sulphide oxidation and carbonate dissolution as a source of CO_2 over geological timescales. *Nature*, **507**(7492), 346–349.

Trillo I., Jenkins T., Redmon D., Hilgart T. and Trillo J. (2004). Implementation of feedforward aeration control using on-line offgas analysis: the Grafton WWTP experience. *Proceedings of the Water Environment Federation*, **2004**, 27–45.

Van der Heyden C., Vanthillo B., Pieters J. G., Demeyer P. and Volcke E. I. P. (2016). Mechanistic modeling of pollutant removal, temperature, and evaporation in chemical air scrubbers. *Chemical Engineering & Technology*, **39**(10), 1785–1796.

Van der Lans R. (2000). Gas-liquid interphase transport. Advanced course on environmental biotechnology. In: R. Van der Lans (ed.), TU Delft and University of Leiden, Delft, Leiden, The Netherlands, pp. 16–17.

van Langerak E. P. A., Beekmans M. M. H., Beun J. J., Hamelers H. V. M. and Lettinga G. (1999). Influence of phosphate and iron on the extent of calcium carbonate precipitation during anaerobic digestion. *Journal of Chemical Technology & Biotechnology*, **74**(11), 1030–1036.

van Rensburg P., Musvoto E. V., Wentzel M. C. and Ekama G. A. (2003). Modelling multiple mineral precipitation in anaerobic digester liquor. *Water Research*, **37**(13), 3087–3097.

Vaneeckhaute C., Claeys F. H. A., Tack F. M. G., Meers E., Belia E. and Vanrolleghem P. A. (2018). Development, implementation, and validation of a generic nutrient recovery model (NRM) library. *Environmental Modelling & Software*, **99**, 170–209.

Varga E., Hauduc H., Barnard J., Dunlap P., Jimenez J., Menniti A., Schauer P., Lopez Vazquez C. M., Gu A. Z., Sperandio M. and Takács I. (2018). Recent advances in bio-P modelling – a new approach verified by full-scale observations. *Water Science and Technology*, **78**(10), 2119–2130.

Wang G. Q., Yuan X. G. and Yu K. T. (2005). Review of mass-transfer correlations for packed columns. *Industrial & Engineering Chemistry Research*, **44**(23), 8715–8729.

Wang J., Burken J. G. and Zhang X. (2006). Effect of seeding materials and mixing strength on struvite precipitation. *Water Environment Research*, **78**(2), 125–132.

Wang R., Wilfert P., Dugulan I., Goubitz K., Korving L., Witkamp G.-J. and van Loosdrecht M. C. M. (2019). Fe(III) reduction and vivianite formation in activated sludge. *Separation and Purification Technology*, **220**, 126–135.

Wilfert P., Mandalidis A., Dugulan A. I., Goubitz K., Korving L., Temmink H., Witkamp G. J. and Van Loosdrecht M. C. M. (2016). Vivianite as an important iron phosphate precipitate in sewage treatment plants. *Water Research*, **104**, 449–460.

Wilfert P., Meerdink J., Degaga B., Temmink H., Korving L., Witkamp G. J., Goubitz K. and van Loosdrecht M. C. M. (2020). Sulfide induced phosphate release from iron phosphates and its potential for phosphate recovery. *Water Research*, **171**, 115389.

Zhang Y., Piccard S. and Zhou W. (2015). Improved ADM1 model for anaerobic digestion process considering physico-chemical reactions. *Bioresource Technology*, **196**, 279–289.

Zhang L., Qiu Y.-Y., Zhou Y., Chen G.-H., van Loosdrecht M. C. M. and Jiang F. (2021). Elemental sulfur as electron donor and/or acceptor: Mechanisms, applications and perspectives for biological water and wastewater treatment. *Water Research*, **202**, 117373.

Zhu Z., Yogev U., Keesman K. J. and Gross A. (2021). Onsite anaerobic treatment of aquaponics lettuce waste: digestion efficiency and nutrient recovery. *Aquaculture International*, **29**(1), 57–73.

IWA Publishing's authorised EU representative for General Product
Safety Regulations is Diane D'Arras, 15 rue Duret, 75116 Paris,
France, e-mail: safety@iwap.co.uk.

Printed and bound by CPI Group (UK) Ltd, Croydon, CR0 4YY
12/05/2026
02108863-0017